Maria Borcsa • Peter Rober

Editors

Research Perspectives in Couple Therapy

Discursive Qualitative Methods

 Springer

Editors
Maria Borcsa
Institute of Social Medicine
Rehabilitation Sciences and Healthcare
  Research
University of Applied Sciences Nordhausen
Nordhausen, Germany

Peter Rober
Institute for Family & Sexuality Studies
University of Leuven
Leuven, Belgium

ISBN 978-3-319-23305-5   ISBN 978-3-319-23306-2   (eBook)
DOI 10.1007/978-3-319-23306-2

Library of Congress Control Number: 2015954656

Springer Cham Heidelberg New York Dordrecht London

Printed on acid-free paper

Springer International Publishing AG Switzerland is part of Springer Science+Business Media
(www.springer.com)

Foreword: An Invitation to Enter a Qualitative Research Dialogue

Two young people from different European countries meet, fall in love, and decide to live together. After a while, tensions develop in their relationship. They visit a therapist who specializes in helping couples. Their therapy sessions are recorded on video. Sometime afterwards, with their permission, what was said in each session was transcribed and analyzed by researchers interested in understanding more about how this kind of therapy works. That is what this book is about.

How can we begin to make sense of what these researchers have done? I suggest that there are perhaps at least four perspectives that are relevant: personal, esthetic, scientific, and practical.

From a personal perspective, anyone reading these chapters already has their own way of making sense of love, living together, cultural difference, and therapy. In the light of this personal knowledge, the central question that can be asked about the analyses offered in this book is: "how does this take forward, extend, or deepen my appreciation of these aspects of the world in which I live?" It may be useful, therefore, before starting to read this book, to take a few moments to engage in personal reflection on what you already know and believe in respect of these matters, and what you might expect these researchers to find. This process can be facilitated by consulting the brief account of the project provided at the beginning of the second chapter, then pausing, to collect your own thoughts. I believe that the authors would appreciate this kind of thoughtful and reflective approach to their work. It seemed to me, reading their contributions, that none of them are trying to persuade. Rather, they are in the business of offering.

Each chapter offers an assemblage of different threads of understanding and experience, based on a brief, clearly articulated statement of a particular theoretical perspective. Each chapter also includes careful selection and highlighting of examples of moments within the therapy that have particular significance. Woven around these segments of interaction is an invitation to make sense of them in a particular way. In engaging with this kind of work, it can be useful to think about art: each chapter as an artwork, editors as curators, the book as an exhibition, readers as visitors. What is on offer is a "way of seeing" (Berger, 1973).

At the same time, the work in this book draws on the concepts and methods of social science and psychology, specifically from within the field of psychotherapy research. Readers who are interested in the technical details of the methodologies that are being applied can consult references that are provided within each chapter. The work that is reported here reinforces the importance of the researcher as "bricoleur" (Kincheloe, 2005). Faced with the challenge of making sense of the complexity of interactions between client and therapist, these researchers have found it necessary to move beyond standard methodologies, to develop new ways of finding meaning.

In the end, research into psychotherapy is always viewed in relation to the degree to which it contributes to practice. The authors whose work is represented in this book are all practicing psychotherapists. The research that they report has emerged out of practice and is closely tied into the continued development of practice. Furthermore, they are practitioner–researchers whose ideas and therapy approaches form part of a network. They meet together to learn from each other, at conferences and workshops. The book as a whole can therefore be regarded as comprising a documentation of the lived experience of learning that occurs within a community of practice (Wenger, 1998). This book highlights the research face of that community. Many other publications, indicated within each chapter, highlight the practical or clinical face.

The concept of *dialogue* comprises a constant thread through the studies that are reported in this volume. What does an authentic dialogue look like, within a therapy session? What are the conditions under which it takes place? How does it help? At another level, the book as a whole is a dialogue, between alternative perspectives on these issues. As readers, we are invited not only to enter this conversation, but also to continue it, in our personal and professional lives.

Oslo, Norway   John McLeod
                                                          Department of Psychology
                                                              University of Oslo

References

Berger, J. (1973). *Ways of seeing*. Harmondsworth: Penguin.
Kincheloe, J. L. (2005). On to the next level: Continuing the conceptualization of the bricolage. *Qualitative Inquiry, 11*, 323–350.
Wenger, E. (1998). *Communities of practice: Learning, meaning, and identity*. Cambridge: Cambridge University Press.

Preface

It was at an international seminar at Leuven University in October 2009 when the very first research on the case of Victoria and Alfonso started. Some of us met for three days in an old thirteenth century abbey, watched the videotapes of the four sessions of the therapy, and talked about them. This group did not know at that time where they were going with this investigation, but the conversations were fascinating and all felt inspired, especially through the different views, different observations, different questions, and different interpretations. Everyone was impressed by the myriad of facets of the case and by the touching aliveness of the interactions between both the partners and the therapists. This was the beginning of the idea to search for ways of doing systematic research in which this richness would be respected. Already during this first encounter, it was obvious that this was going to be a challenge because of the complexity of multi-actor dialogues.

One year later, the *European Family Therapy Research Group* (EFTRG) was founded at a meeting at Jyväskylä University (Finland). We were therapists/researchers from Finland, Portugal, Italy, Germany, Belgium, and Greece. The decision to meet a few times a year in order to support and inspire each other in doing research in the field of marital and family therapy went along with the choice to start our joint work on the case of Victoria and Alfonso.

In the following years, EFTRG had meetings in Helsinki (2011), Thessaloniki, Nottingham, and Porto (2012), Istanbul (2013), Copenhagen, Heidelberg, and Crete (2014). We presented papers on our research of Alfonso's and Victoria's therapy at different conferences (1st International Conference on Dialogical Practices Helsinki 2011; QRMH4 Nottingham 2012; SPR Porto 2012; EFTA Conference Istanbul 2013; SPR Copenhagen 2014; 1st European Systemic Research Conference Heidelberg 2014; QRMH5 Crete 2014) and, more and more, the idea took shape that we should accumulate a written report about the different approaches we had developed in doing discursive research on this couple therapy. This book is the result.

Our gratitude goes to the clients, who we named Victoria and Alfonso, as well as their therapists. First and foremost, it was them who gave us the opportunity to have a close look at one therapeutic system from different angles.

Furthermore, we thank the organizers as well as the scientific committees of the diverse conferences where we presented, as they encouraged our project by providing us with time and space to discuss the ongoing work with different audiences.

We would also like to thank John Shotter and John McLeod for their support and inspiration, Anna Ptitsyna for her valuable English language checks, and Manuela Günther, who meticulously made the editorial corrections.

Last, but not least, we thank Jennifer Hadley from Springer International for her patience.

Nordhausen, Germany   Maria Borcsa
Leuven, Belgium   Peter Rober

Contents

About the Contributors

Evrinomy Avdi, Ph.D. is Assistant Professor of Clinical Psychology at the School of Psychology of the Aristotle University of Thessaloniki, Greece. She is a clinical psychologist, dramatherapist, and psychodynamic psychotherapist. Her research interests lie in applying discursive and narrative research to the study of therapy process, as well as the experience of serious illness. She is particularly interested in exploring the links between deconstructive research and actual clinical practice.

Maria Borcsa, Ph.D. is Professor of Clinical Psychology at the University of Applied Sciences Nordhausen, Germany. She is an accredited psychological psychotherapist, family and systemic therapist, supervisor, and trainer. She is a founding member and member of the executive committee of the Institute of Social Medicine, Rehabilitation Sciences and Healthcare Research at UAS Nordhausen. For many years, she has served as a board member of the German Systemic Society (Systemische Gesellschaft) as well as of the European Family Therapy Association (EFTA). Currently, she holds the position as the President of EFTA.

Lisa Fellin, Ph.D. is Senior Lecturer in Psychology at the University of East London, UK. She holds a PhD in Clinical Psychology and is an accredited systemic psychotherapist. She is currently the Research Director for the Professional Doctorate in Counselling Psychology and the Module coordinator for Family & Systemic Therapies on the MSc in Clinical & Community Psychology, as well as a trainer at the European Institute of Systemic-relational Therapies in Milan, Italy. She is a member of the European Family Therapy Association (EFTA) and of the UK Association of Family Therapy and Systemic Practice (AFT).

Juha Holma, Ph.D. is Lecturer at the Department of Psychology, University of Jyväskylä, Finland. He is an accredited psychotherapist, family therapist, supervisor, and trainer. He has worked as a group facilitator in the treatment groups for men who have battered their partners, and he is the responsible researcher in a project on group treatment for batterers and on couple treatment in intimate partner violence.

Aarno Laitila is Professor of Psychology at University of Eastern Finland, Joensuu, Finland. He is an accredited psychologist, psychotherapist, and psychotherapy trainer as well as supervisor, specialized in family therapy. He is one of the founding members of Central Finland Association of Family Therapy Trainers and a present member of the Finnish National Consortium of Universities for Psychotherapist Training.

Mary Olson, Ph.D. is Assistant Professor of Psychiatry at the University of Massachusetts Medical School and Adjunct Professor at the Smith College School for Social Work. A teacher, researcher, and practitioner, she founded a training facility, the Institute for Dialogic Practice, in Haydenville, Massachusetts (USA) and has specialized in developing new training methods for dialogic practice: a "post-systems" approach. Dr. Olson was a Senior Fulbright Scholar (2001–2002) to Finland in the Department of Clinical Psychology at the University of Jyväskylä. A member of the American Academy of Family Therapists, she maintains a private practice and treats and consults with individuals, couples, and families on a variety of clinical issues.

Helena Päivinen, M.A. is a doctoral student at the Department of Psychology, University of Jyväskylä, Finland. Her dissertation concerns positioning as a discursive phenomenon and a tool in psychotherapy. She is a clinical psychologist specialized in intimate partner violence issues.

Peter Rober, Ph.D. is a clinical psychologist, family therapist, and family therapy trainer at Context -Center for marital and family therapy (UPC KU Leuven, Belgium). He is Professor of Family Therapy at the Institute for Family and Sexuality Studies (Medical School of K.U. Leuven, Belgium). His research interests focus on what is said and not said in families and in family therapy. He has a special interest in the process in family therapy, in particular in the use the therapist's self and in the therapist's inner conversation.

Jaakko Seikkula, Ph.D. is Professor of Psychotherapy at University of Jyväskylä, Finland. He is a psychotherapy trainer and has been involved in developing Open Dialogue approach in most severe psychiatric crises and in conducting several studies of the outcome and dialogical processes of the Open Dialogue approach in psychosis and couple therapy for depression. He is one of the founding members of the International Network in Treatment of Psychosis and the European Family Therapy Research Group.

Valeria Ugazio, Ph.D. is Professor of Clinical Psychology at University of Bergamo, Italy. She is an accredited systemic psychotherapist, supervisor, and trainer. She is the founder and scientific director of the European Institute of Systemic-relational Therapies in Milano, where she runs 4-year training courses in systemic family, couple, and individual psychotherapy for psychologists and medical doctors and practices psychotherapy.

Jarl Wahlström, Ph.D. is Professor of Clinical Psychology and Psychotherapy and director of the Psychotherapy Training and Research Centre at the Department of Psychology, University of Jyväskylä, Finland. He is head of the national integrative postgraduate specialization program in psychotherapy for psychologists, organized by the Finnish University Network in Psychology, Psykonet. He has an advanced training in family and systems therapy and more than 30 years' experience as a clinical psychologist, family therapist, trainer and consultant, and university teacher. His main research interest lies in psychotherapy discourse, a topic on which he has published in several international journals and books. He is the supervisor of several doctoral dissertations.

Chapter 1
The Challenge: Tailoring Qualitative Process Research Methods for the Study of Marital and Family Therapeutic Sessions

Peter Rober and Maria Borcsa

In this book, we present different qualitative research (QR) approaches to study marital and family therapy (MFT) sessions. In several ways this is unusual. Up to now, although there has been increased interest in looking at psychotherapy in terms of discourse, narrative and dialogue, most psychotherapy research is still quantitative and psychotherapy research in the field of MFT is rather scarce. Furthermore, while QR is appreciated in the fields of sociology, pedagogy and anthropology, for instance, in the fields of clinical psychology and psychotherapy it has just started to be accepted as a valid kind of research. This is an especially important development as words and meanings—the area qualitative research traditionally focuses on—are central to psychotherapy in general and to MFT in particular, and qualitative methods can bring research closer to the complexity of MFT practice.

Clinicians and Researchers

Many practitioners have not found empirical research on psychotherapy very valuable for them (Lambert, 2013a). Research was mostly focussed on outcome and used quantitative measurements. Indeed, most money and efforts of research in the field of psychotherapy went into studying Randomised Controlled Trials (RCT) (Lambert, 2013a). Such research is valuable for academics and policy makers, as it helps to determine in broad strokes that psychotherapy works. For clinicians,

P. Rober (✉)
Department of Neurosciences, Institute for Family and Sexuality Studies, KU Leuven and Context, UPC KU Leuven, Belgium
e-mail: peter.rober@med.kuleuven.be

M. Borcsa
Institute of Social Medicine, Rehabilitation Sciences and Healthcare Research, University of Applied Sciences Nordhausen, Nordhausen, Germany

© Springer International Publishing Switzerland 2016
M. Borcsa, P. Rober (eds.), *Research Perspectives in Couple Therapy*,
DOI 10.1007/978-3-319-23306-2_1

however, such research is often less useful, as it does not contribute to new perspectives or ideas in the difficult situations clinicians are confronted with in their practices. Furthermore, RCT research evoked the expectation that it would help to determine which of the psychotherapy approaches worked better than other approaches. Yet, by now RCT has shown that all approaches have impacts and that the differences between them in terms of effectiveness are minimal (Lambert, 2013b; Sprenkle & Blow, 2004; Wampold, 2001). While from a certain academic perspective it is disappointing that RCT research did not find differences between approaches, still RCT studies were important as they showed psychotherapy effectiveness in general (Hubble, Duncan, & Miller, 1999). However, based on RCT research it is not possible to determine *what* makes psychotherapy work (Duncan, Miller, Wampold, & Hubble, 2010). And the fact that all approaches are effective suggests that probably common factors are the most potent ones in the therapeutic process (Hubble et al., 1999; Sprenkle, Davis, & Lebow, 2009).

Against this background, the realisation began to grow that we need more process research in order to understand what makes psychotherapy work (Lambert, 2013b). Furthermore, it is obvious that qualitative research methods are needed in addition to quantitative methods in order to study the therapy process in detail. Prominent authors speak of the coming of age of QR in the field of psychotherapy research (McLeod, 2013, p. 49).

Process research is probably more appealing for clinicians than RCT research, as it comes much closer to the unpredictable and ambiguous process of therapy that clinicians experience in their day-to-day practice. The unique story of the client, with his/her pain and suffering that mobilises therapists to give it their best shot to be helpful, resonates in this kind of research that strives to be faithful to the client's voice. This is important, especially in the field of MFT, where there has always been a strong focus on therapists in their daily practices rather than on research or theory (Sprenkle & Piercy, 2005). First there was practice, and later theory and research were added. Indeed, the field started when different clinicians in different places in Europe and the US started to work with families rather than with individual patients. They could be called "researcher–clinicians", as they studied—often in their teams—the impact of their own interventions, with the aid of video recordings or one-way-screens (Sprenkle & Piercy, 2005). The field grew because these charismatic pioneers inspired therapists with their ideas about the importance of families and interactional patterns. Models were developed based on clinical experience with families, and schools were established. During the first decades, theory and research served especially to legitimate the clinical expertise of the pioneers. When later MFT research—independent of the main family therapy schools—began to blossom, this resulted in a research-practice gap, in which the clinicians often feel unappreciated and misunderstood by the researchers, and in which the researchers disdain the clinicians and underestimate the importance of the experience, the intuition and the wisdom of the latter. Clinicians dismiss research as irrelevant and difficult to understand, while researchers fail to understand that they can only do research if resourceful and creative clinicians come up with new ideas of new techniques that they can test later for their effectiveness.

Experience and Process

While the gap between clinical practice and research has widened due to the strong emphasis on evidence-based approaches and RCT research (Lambert, 2004), there are trends in the field of psychotherapy research that show promise in their contributing to closing the gap. One of these trends is the research on monitoring and feedback informed clinical work (Lambert, 2010). Another trend is the growing interest in QR on the process of therapy (McLeod, 2013).

QR can have at least two foci in the field of psychotherapy: experience and process of therapy (McLeod, 2013). Most of the research of the first category is concerned with how clients experience certain aspects of therapy (e.g. the therapeutic alliance, the process, etc.) (e.g. Hartzell, Seikkula, & Von Knorring, 2009; Olson & Russell, 2004). However, there is also research into the therapist's experiences (e.g. Goddard, Murray, & Simpson, 2008) and into the experience of the family members (e.g. Roberts, 1996; Sampson, 2013; Sheridan, Peterson, & Rosen, 2010). For these studies, participants (clients, therapists or family members) are interviewed to self-report on their experiences in therapy. Sometimes special data-collection procedures are used, like tape-assisted recall procedures (e.g. Rober, Elliott, Buysse, Loots, & De Corte, 2008). Also for data analysis, specific methodological procedures are introduced like Grounded Theory (Charmaz, 2006) or Interpretative Phenomenological Analysis (Smith, Flowers, & Larkin, 2009).

Besides research into the experience of therapy, there is also research on the process of therapy. In these kinds of research studies, it is as if you look at psychotherapy with a microscope (McLeod, 2013). The objective of the researcher is to observe and analyse rigorously and systematically what happens in the therapy session, in order to notice what would have remained unnoticed if the researcher had not looked so systematically. The aim is to describe as faithfully as possible what the researcher has observed. This describing usually entails the highlighting of patterns and implicit structures, as well as the careful choice of words to respect as much as possible the complexity and richness of the session. Raw data of this approach usually consist of audio or videotape recordings that are transcribed using certain notation methods that are usually based on the method originally developed by Gail Jefferson (e.g. Jefferson, 2004). Most often the data are then analysed using some form of discursive method, like conversation analysis (CA) or discourse analysis (DA), or by a narrative analytic (NA) method. CA, DA and NA all have a social construction rather than a realist philosophical foundation, in the sense that they view language as a social construction produced by people in a certain social, cultural and historical context (Gergen, 1999). Notwithstanding this common philosophical foundation, CA, DA and NA can be distinguished from one another in the way they emphasise different aspects of social construction in their practical application.

A Closer Look at CA, DA and NA

While CA and DA both exist in different shapes and forms (e.g. poststructuralist DA, discursive DA and so on), it is not so easy to define exactly what the difference between CA and DA is. Let us look at these approaches in more detail. We will characterise each of them and give one example of a study that illustrates what is typical about this approach.

1. Conversation analysis: To put it (too) simply, we could say that conversation analysis focuses mainly on "talk-in-interaction". In CA, there is an emphasis on the way interlocutors accomplish, through their language—both verbal and non-verbal—on a turn-to-turn basis, the production of interactional regularities like alliance (Sutherland & Couture, 2007); response to blame (Friedlander, Heatherington, & Marrs, 2000); collaboration (Sutherland & Strong, 2011); a therapist's linguistic strategies (Gale & Newfield, 1992) and so on. A good example of CA research in MFT is Shari Couture's and Olga Sutherland's study of a Karl Tomm session and more specifically of Tomm's advice-giving (Couture & Sutherland, 2006). In their study, they focus on advice as an interactional accomplishment to which both clients and therapists contribute rather than a unilateral act administered by the counsellor to the client. They describe the communicative processes and strategies involved in the offering of advice in the context of family therapy. They show that advice-giving does not necessarily involve an all-knowing therapist enlightening the family about the "right things to do". Their analysis reveals that advice-giving can also be a dialogical, cyclical, stepwise process of opening space for new possibilities and forward-moving orientations for the family.

2. Discourse analysis: in contrast with CA, in discourse analysis the focus is not only on what happens between the interlocutors, but also on external sociocultural power relations that invade and shape the interlocutors' interactions. So the DA researcher is interested in the way societal discourses around health (Avdi, 2005), culture (Singh, 2009), gender and so on might influence the way that human interactions and language take shape in a particular situation. Karatza and Avdi (2011) is a good example of the use of DA to study the process of psychotherapy. The paper is aimed at examining the process of therapy with families in which one member has a diagnosis of psychosis and focusses on shifts in the subject positions occupied by the identified patient in the clinical dialogue. The study demonstrates that DA provides a useful method for the detailed study of the ways in which meanings evolve within the clinical dialogue, as it zooms in on collapses of meaning and narrative, and on alienation from shared communicative practices, in psychosis. The study suggests that an important aspect in clinical work with these families relates to the repositioning of the patient and on facilitating flexible, agentic and reflexive subject positions.

   In the study of Karatza and Avdi, the concept of positioning is central (Davies & Harré, 1990; Harré, Moghaddam, Pilkerton Cairnie, Rothbart, & Sabat, 2009; Harré & Van Langenhove, 1999). More dynamic than the sociological concept of *role*, it points to the rights and duties of a person in a certain social interaction. It refers to the meanings that people discern in the actions of others and that they give to what they do themselves as normative constraints or opportunities for action in a certain local conversation. The concept of positioning enables DA to spell out some of the ways in which clients (and sometimes therapists) are trapped in constraining discursive contexts that limit their possibilities for action. It also allows DA to describe ways in which therapists can offer their clients alternative paths with a broader range of choices.

3. Narrative analysis: besides CA and DA, narrative analysis (NA) also can be used to qualitatively study a psychotherapy process (McLeod, 2013). In NA the focus is on the client's talk as a kind of storytelling, in which the client gives meaning to her/his lived experience. The narrative researcher is interested in the way the client's story unfolds in the session and in the contribution of the therapist to the narrative. A good example of this kind of research can be found in the study of Rober, van Eesbeek, and Elliot (2006) in which they examined the process of storytelling about domestic violence in a family therapy session. The analysis made it possible to describe storytelling in family therapy as a collaborative, dialogical process, accomplished by all participants in the conversation. The recounting of stories of violence seemed to go hand in hand with more repressive modes of interaction that discouraged the telling of these stories. In the back-and-forth process between voices of hesitation and voices of reassurance, the level of safety in the session is weighed by the participants. In so far as the voices of hesitation can be reassured of the safety, it becomes gradually possible to talk about delicate, problematic experiences such as violence in the family.

The Specificity of an MFT Setting and Research

What the study of Rober et al. (2006) also illustrates is that in order to investigate complex interactions like family therapy sessions it is necessary to be methodologically open-minded and creative. Their study combined narrative analysis with elements from CA and tape-assisted recall. This helped them to find richer answers to their questions: while the narrative analysis gave the researchers an overall focus on storytelling, using CA tools helped them to move beyond the content of the stories and make the process of storytelling accessible for analysis. An additional perspective on the session was offered by the tape-assisted recall procedure, in the sense that it gave access to what the therapist had not said, but what was present in the therapist's inner conversation.

Different authors (e.g. Gale, Odell, & Nagireddy, 1995) have written about the way research can be enriched by the combination of methods, what McLeod (2001) calls "bricolage", Tseliou (2013) "mixed types of analysis" or Flick (2009) "triangulation". So researchers can combine CA and DA (e.g. Kogan, 1998), or combine CA, DA and narratology (e.g. Kurri & Wahlström, 2005). Or they can do Grounded Theory in a novel way that better fits the specificity of their research question or the specificity of their participants (e.g. Frosh, Burck, Strickland-Clark, & Morgan, 1996). Very often researchers in the field of MFT will have to resort to *bricolage* because of the complexity—be it on the content or on the structural level—of their specific therapy settings. On the content level, multi-actor settings like couple sessions or family sessions are usually emotionally intense, sometimes conflictual, and family members differ in their motivations to go into therapy, in so far that some family members even may not agree that therapy is needed. On the structural level, the researcher has to face a couple or family system with their history of interaction as well as the new established therapeutic system which is the family and the therapist. Such tension-filled and multilayered settings demand specific methodological tools in order to make research possible.

Narrative and Dialogical Conceptual Frames

The complexity of multi-actor settings does not only demand new and complex methods, but it also calls for new and complex conceptual frames that better fit the specificity of the setting.

An important conceptual frame for a lot of researchers in the field of MFT is the narrative frame (e.g. Avdi, 2005). The narrative perspective has enriched the field's view of the family therapeutic encounter in profound ways, as it helped researchers to better understand how people make sense of their experiences and give meaning to their life through stories. McAdams (1997), for instance, makes a link between human identity and narrative as he argues that people "are" the stories they tell. A person's identity itself takes the form of an inner story, complete with settings, scenes, character, plot and themes, providing coherence and purpose to one's life (McAdams & Janis, 2004). Narrative ideas like these have been important for our field as they inspired researchers to use narrative metaphors like story and authoring to analyse the processes that take place in MFT sessions and to describe in detail how identities are socially constructed and reconstructed by the stories that are told and retold in the session.

Another conceptual frame that is important in the field of MFT qualitative research is the dialogical perspective. This perspective puts the spotlight not on the stories themselves, but rather on storytelling as dialogue (e.g. Rober et al., 2006) and on narratives in action (Wortham, 2001). While stories may define who the narrator is, in the telling of the story the relationship between the narrator and his/her audience emerges. In this dialogical approach, storytelling is seen as a performance in context, implying that stories only exist through the presence of others who listen to these stories. Referring to some of Bakhtin's ideas (Bakhtin, 1981, 1984, 1986), storytelling can be conceptualised as a dialogical phenomenon: Stories are told using cultural tools like words, expressions and speech genres, and the storyteller has the listener constantly in mind.

For researchers, deeper reflection on the dialogical aspects of storytelling is a useful addition to a narrative view, as storytelling in a multi-actor setting is particularly complex. For instance, the stories told in a family session develop in the dialogical interaction and—as such—did not exist beforehand. Storytelling is not the simple result of the storyteller's pre-existing intention to share his/her experiences. Rather, inspired by Bakhtin (1981), we can describe a family session as a forum where stories develop gradually and in unpredictable ways out of the contradiction-ridden and tension-filled interaction of all the interlocutors present. Every utterance is implicitly or explicitly evaluated by others and their verbal and nonverbal reactions invite new utterances in a complex dialectical dance of differences and similarities (Baxter, 2004).

The Challenge

While the complexity of conceptual frames like the narrative perspective and Bakhtin's philosophy has a good fit with the interactional complexity of the MFT setting as a co-constructive and responsive process between people, it is clear that it remains a challenge to use these (or other) conceptual frameworks in researching multi-actor settings. There is, for instance, some research on the dialogical self that can be inspiring for family therapists (e.g. Leiman,2004; Salgado, Cunha, & Bento,2013), but going from studying the dialogical construction of the self to studying a couple or family session as a dialogue is still a big leap. In fact, this is the challenge to which this book tries to formulate some answers: *In which ways can we do research on the process of MFT, respecting the specificity of the setting?*

The different contributors in this volume will each explain how they tried to answer this question by outlining their theoretical background, explaining their methodological approach and illustrating the latter by using it to study the therapy of Victoria and Alfonso.

Therefore, let us now look at the analysed case.

References

Avdi, E. (2005). Negotiating a pathological identity in the clinical dialogue: Discourse analysis of a family therapy. *Psychology and Psychotherapy: Theory, Research and Practice, 78,* 493–511.

Bakhtin, M. (1981). *The dialogic imagination.* Austin, TX: University of Texas Press.

Bakhtin, M. (1984). *Problems of Dostoevsky's poetics.* Minneapolis, MN: University of Minneapolis Press.

Bakhtin, M. (1986). *Speech genres & other late essays.* Austin, TX: University of Texas Press.

Baxter, L. A. (2004). Dialogues of relating. In R. Anderson, L. A. Baxter, & K. N. Cissna (Eds.), *Dialogues: Theorizing difference in communication studies* (pp. 107–124). London: Sage.

Charmaz, K. (2006). *Constructing grounded theory.* Thousand Oaks, CA: Sage.

Couture, S. J., & Sutherland, O. (2006). Giving advice on advice giving: A conversation analysis of Karl Tomm's practice. *Journal of Marital and Family Therapy, 32,* 329–344.

Davies, B., & Harré, R. (1990). Positioning: The discursive production of selves. *Journal for the Theory of Social Behaviour, 20*, 43–63.

Duncan, B. L., Miller, S. D., Wampold, B. E., & Hubble, M. A. (Eds.). (2010). *The heart and soul of change: Delivering what works in therapy* (2nd ed.). Washington, DC: APA.

Flick, U. (2009). *An introduction to qualitative research* (3rd ed.). London, UK/Thousand Oaks, CA: Sage.

Friedlander, M. L., Heatherington, L., & Marrs, A. L. (2000). Responding to blame in family therapy: A constructionist/narrative perspective. *American Journal of Family Therapy, 28*, 133–146.

Frosh, S., Burck, C., Strickland-Clark, L., & Morgan, K. (1996). Engaging with change: A process study of family therapy. *Journal of Family Therapy, 18*, 141–161.

Gale, J., & Newfield, N. (1992). A conversation analysis of a solution focused marital therapy session. *Journal of Marital and Family Therapy, 18*, 153–165.

Gale, J. E., Odell, M., & Nagireddy, C. S. (1995). Marital therapy and self-reflexive research: Research and/or intervention. In G. H. Morris & R. J. Chenail (Eds.), *The talk of the clinic: Explorations in the analysis of medical and therapeutic discourse* (pp. 105–129). Hillsdale, NJ: Lawrence Erlbaum.

Gergen, K. J. (1999). *An invitation to social construction*. London: Sage.

Goddard, A., Murray, C. D., & Simpson, J. (2008). Informed consent and psychotherapy: An interpretative phenomenological analysis of therapists' views. *Psychology and Psychotherapy: Theory, Research and Practice, 81*, 177–191.

Harré, R., Moghaddam, F. M., Pilkerton Cairnie, T., Rothbart, D., & Sabat, S. R. (2009). Recent advances in positioning theory. *Theory & Psychology, 19*, 5–31.

Harré, R., & Van Langenhove, L. (Eds.). (1999). *Positioning theory*. Oxford: Blackwell.

Hartzell, M., Seikkula, J., & von Knorring, A.-L. (2009). What children feel about their first encounter with Child and Adolescent Psychiatry. *Contemporary Family Therapy, 31*, 177–192.

Hubble, M. A., Duncan, B. L., & Miller, S. (Eds.). (1999). *The heart and soul of change*. Washington, DC: American Psychological Association.

Jefferson, G. (2004). Glossary of transcript symbols with an Introduction. In G. H. Lerner (Ed.), *Conversation analysis: Studies from the first generation* (pp. 13–23). Philadelphia, PA: John Benjamins.

Karatza, H., & Avdi, E. (2011). Shifts in subjectivity during the therapy for psychosis. *Psychology and Psychotherapy: Theory, Research and Practice, 84*, 214–229.

Kogan, S. M. (1998). The politics of making meaning: Discourse analysis or a 'postmodern' interview. *Journal of Family Therapy, 20*, 229–251.

Kurri, K., & Wahlström, J. (2005). Placement of responsibility and moral reasoning in couple therapy. *Journal of Family Therapy, 27*, 352–369.

Lambert, M. J. (2004). Introduction and historical overview. In M. J. Lambert (Ed.), *Bergin & Garfield's handbook of psychotherapy and behavior change* (5th ed., pp. 3–15). New York: Wiley.

Lambert, M. J. (2010). Yes, it is time for clinicians to routinely monitor treatment outcome. In B. L. Duncan, S. D. Miller, B. E. Wampold, & M. A. Hubble (Eds.), *The heart and soul of change: Delivering what works in therapy* (2nd ed., pp. 239–266). Washington, DC: APA.

Lambert, M. J. (2013a). Introduction and historical overview. In M. J. Lambert (Ed.), *Bergin & Garfield's handbook of psychotherapy and behavior change* (6th ed., pp. 3–20). New York: Wiley.

Lambert, M. J. (2013b). The efficacy and effectiveness of psychotherapy. In M. J. Lambert (Ed.), *Bergin & Garfield's handbook of psychotherapy and behavior change* (6th ed., pp. 169–218). New York: Wiley.

Leiman, M. (2004). Dialogical sequence analysis. In H. Hermans & G. Dimaggio (Eds.), *The dialogical self in psychotherapy* (pp. 255–270). Hove: Brunner-Routledge.

McAdams, D. P. (1997). *The stories we live by: Personal myths and the making of the self*. New York: Guilford.

McAdams, D. P., & Janis, L. (2004). Narrative identity and self-multiplicity. In L. E. Angus & J. McLeod (Eds.), *The handbook of narrative and psychotherapy: Practice, theory and research* (pp. 159–173). London: Sage.

McLeod, J. (2001). *Qualitative research in counselling and psychotherapy*. London: Sage.

McLeod, J. (2013). Qualitative research: Methods and contributions. In M. J. Lambert (Ed.), *Bergin & Garfield's handbook of psychotherapy and behavior change* (6th ed., pp. 49–84). New York: Wiley.

Olson, M., & Russell, C. (2004). Understanding change in conjoint psychotherapy: Inviting clients to comment upon the validity of standardized change scores. *Contemporary Family Therapy, 26*, 261–278.

Rober, P., Elliott, R., Buysse, A., Loots, G., & De Corte, K. (2008). What's on the therapist's mind? A grounded theory analysis of family therapist reflections during individual therapy sessions. *Psychotherapy Research, 18*, 48–57.

Rober, P., van Eesbeek, D., & Elliot, R. (2006). Talking about violence: A microanalysis of narrative processes in a family therapy session. *Journal of Marital and Family Therapy, 32*, 313–328.

Roberts, J. (1996). Perceptions of the significant other of the effects of psychodynamic psychotherapy, implications for thinking about psycho-dynamic and systems approaches. *British Journal of Psychiatry, 168*, 87–93.

Salgado, J., Cunha, C., & Bento, T. (2013). Positioning microanalysis: Studying the self through the exploration of dialogical processes. *Integrative Psychological and Behavioral Science, 47*, 325–335.

Sampson, J. (2013). The lived experience of family members of persons who compulsively hoard: A qualitative study. *Journal of Marital and Family Therapy, 39*, 388–402.

Sheridan, M., Peterson, B., & Rosen, K. (2010). The experiences of parents of adolescents in family therapy: A qualitative investigation. *Journal of Marital and Family Therapy, 36*, 144–157.

Singh, R. (2009). Constructing 'the family' across culture. *Journal of Family Therapy, 31*, 359–383.

Smith, J., Flowers, P., & Larkin, M. (2009). *Interpretative phenomenological analysis: Theory, method and research*. London: Sage.

Sprenkle, D. H., & Blow, A. J. (2004). Common factors and our sacred models. *Journal of Marital and Family Therapy, 30*, 113–129.

Sprenkle, D. H., Davis, S. D., & Lebow, J. L. (2009). *Common factors in couple and family therapy*. New York: Guilford Press.

Sprenkle, D. H., & Piercy, F. P. (2005). Pluralism, diversity and sophistication in family therapy research. In D. H. Sprenkle & F. P. Piercy (Eds.), *Research methods in family therapy* (2nd ed., pp. 3–18). New York: Guilford Press.

Sutherland, O., & Couture, S. (2007). The discursive performance of the alliance in family therapy: A conversation analytic perspective. *Australian and New Zealand Journal of Family Therapy, 28*, 210–217.

Sutherland, O., & Strong, T. (2011). Therapeutic collaboration: A conversation analysis of constructionist therapy. *Journal of Family Therapy, 33*, 256–278.

Tseliou, E. (2013). A critical methodological review of discourse and conversation analysis studies of family therapy. *Family Process, 52*, 653–672.

Wampold, B. E. (2001). *The great psychotherapy debate: Models, methods and findings*. Mahwah, NJ: Erlbaum.

Wortham, A. (2001). *Narratives in action: A strategy for research and analysis*. New York: Teachers College Press.

Chapter 2
The Couple Therapy of Victoria and Alfonso

Peter Rober and Maria Borcsa

In this chapter, we present the case that will be studied from different angles in this book. We introduce the couple, the context in which the therapy took place and give an overview—session by session—of what happened in the therapy. Throughout the book, the process of therapy will be the background for the different analyses, but only in this chapter the whole development in the four sessions is described. In that way, we try to avoid needless repetitions in the chapters to come. Of course, summarizing a therapeutic process is already an intervention that alters the complexity of what actually happened, as some aspects of the therapy are highlighted while others are not mentioned. So it may already involve a perspective on what is more and what is less important. Still we tried to condense the therapy following central thematic lines in such a way that the reader is familiar with the whole process.

The Couple and the Context

Victoria[1] is a Scandinavian woman aged 25. She contacted the psychotherapy clinic asking help for her and for her partner Alfonso.[2] Alfonso is a Mediterranean man of 21, who came to the country of Victoria for his studies. They have been living together for 3 years.

---

[1] Not her real name.

[2] Not his real name.

P. Rober (✉)
Department of Neurosciences, Institute for Family and Sexuality Studies, KU Leuven and Context, UPC KU Leuven, Belgium
e-mail: peter.rober@med.kuleuven.be

M. Borcsa
Institute of Social Medicine, Rehabilitation Sciences and Healthcare Research, University of Applied Sciences Nordhausen, Nordhausen, Germany

© Springer International Publishing Switzerland 2016
M. Borcsa, P. Rober (eds.), *Research Perspectives in Couple Therapy*,
DOI 10.1007/978-3-319-23306-2_2

The couple therapy took place in a psychotherapy clinic in Victoria's hometown. In this clinic, it is a custom that the clients are asked for written consent for research, as happened also in this case. The sessions were conducted in English (not the first language for any of the participants). There were two therapists present: T1 is an experienced male systemic therapist and T2 (not present in the first session), a younger female trainee in family therapy. A total of four sessions were conducted in a period of 3 months; the interval between the sessions was around 3 weeks.

The sessions were videotaped; the couple gave informed consent to participating in research. The videotapes of the sessions were transcribed. These transcripts were the main data the authors have been working on.[3] During meetings in Belgium (2009), Finland (2011, 2013), Greece (2011), and UK (2012), brief passages of the videotapes were watched by all researchers, so as to give them some sense of the emotional atmosphere of the sessions.

Alliance and Outcome

As it is regular practice in this psychotherapy clinic at the end of every session, the Session Rating Scale (SRS) was completed in order to assess the therapeutic alliance in the meeting (Duncan et al., 2003; Miller, Duncan, Brown, Sparks, & Claud, 2003). Starting from the beginning of the second session, both partners completed further the Outcome Rating Scale (ORS) to evaluate the therapeutic change.

If we want to get a sense of the progress made during therapy in terms of outcome, we have to focus on the ORS scores. The ORS having 25 points altogether from the four subscales is thought to present a life situation without urgent need for change and if the score is below 25, there seems to be need for change. At the start of the therapy, Victoria scored 19 on the ORS, while Alfonso scored 23. In the course of the therapy, Victoria's ORS scores improved from 19 to 37 and Alfonso's from 23 to 32. So the ORS scores seemed to suggest an important improvement for both. Interestingly, for both partners, the most notable change in the ORS scores occurred between the second and the third session (from 23 to 29 for Alfonso; from 19 to 31 for Victoria).

In SRS, out of total 40 points, an optimal evaluation of the session means rating above 36. Victoria's SRS ratings were 36 after every session indicating an optimal evaluation of the alliance. Alfonso's SRS ratings changed from 31 to 33, which are nearly optimal. So both partners experienced the therapeutic alliance as very positive throughout the therapy.

Concluding both the SRS and ORS ratings, it can be noted that the therapy seemed to be helpful for the couple in a therapy process, in which they felt a good working alliance.

---

[3] Except Ugazio & Fellin, Seikkula & Olson and Wahlström, who used also the videotapes as data for their analyses.

The Therapy

When contacting the psychotherapy clinic, Victoria had said the reason for asking for help was Alfonso's difficulties in speaking with her of almost any issues in their lives.

In **the first session**, Victoria explains that she had been depressed for 2 years (following an individual treatment), that she is a lot better now (turn 8), but what happened left scars on their relationship (t10). Alfonso was concerned that he might not have the patience to listen that he used to have anymore (t14), and whenever Victoria wants to talk or asks "is something wrong?" (t38) the tension in the couple rises. Alfonso seems to fear that she might get depressed again, and that he would not be able to be caring like he was the first time (t63). Alfonso thought that maybe he used up a lot of energy supporting Victoria and maybe he needs a rest now (t74). Victoria summarized the problem as "…we can't talk anymore…" (t84). This was difficult because, as Victoria said (t80), "… if there is anything that bothers me I need to share it, I need to talk about it…" That is why they need someone to help them (t82). The therapist explored the difficult situations they were referring to and their cultural differences as well as Alfonso's recent trip to his family in his native country (t129). Victoria explained that it was during a visit to Alfonso's family 2 years ago that she started feeling depressed. So a few months ago, Alfonso went to visit his family again, but Victoria stayed at home, feeling very lonely, however (t137): "I feel no-one while he's there," she says (t139). She expected Alfonso to textmail her during the day, but Alfonso said he was too busy meeting friends and family during the daytime, but he sent a message every evening (t140). This is something Victoria did not understand (t143): "…that's what makes me feel that I'm not your girlfriend" (t145). She explained that she needs some proof that he sometimes thinks of her (t174). When she does not get such proof she feels rejected (t190). They talked about their studies and the history of their relationship up until their trip together to Alfonso's home country. It became clear that Victoria did not feel at ease with Alfonso's family at all: "your family is like the Bold and the Beautiful for me, I'm like trash…" (t331) and that Alfonso' mother "hates me because I stole her son…" (t342). Victoria only feels at ease with Alfonso's brother, who is also the only family member who speaks English. Alfonso translates for her in her contacts with his family.

Almost at the end of the session, the therapist asked about their expectation of the therapy. Victoria referred to a "communication problem" (t390), and Alfonso said that he needed to be able to talk and hoped that Victoria would understand him better. Then the therapist and the couple spoke about the practicalities of their next meetings. At the end of the first session, the therapist gave them a small assignment by asking what they would like to do differently to each other to have a small but visible step towards a better relationship. Victoria asked Alfonso to show some closeness when they have some time together (t462). Alfonso asked Victoria to understand that when he is going some place he is still thinking of her (t476). Then the therapist proposed that Victoria and Alfonso make these small steps in the good direction before their next session (t518).

They came late to **the second session**. When one of the therapists went to meet them in the waiting room, he saw that they were quarrelling and Alfonso had to convince Victoria to come into the office. When they sat down in their chairs, the ORS was filled out. Then the therapist asked Alfonso how the past week had been for him, he said that it had been "all kind of OK." Victoria replied that she felt different. The therapist invited her to say some more about it, and Victoria explained that she did not want to come to the session. She said that she had been working too much and that she was tired and felt sad. When asked about it, she replied: "...like now, when we were coming here, we were talking about the last things that you made us do last time, that we have to make a wish for the other one, and I am sad because I think that it didn't work..." (t33). Alfonso disagreed with her in a comment that was difficult to follow, because he started sentences but did not finish them.

After introducing the co-therapist, therapist 1 tried to continue discussing what Alfonso had meant, but Alfonso could not give a clear expression to his thoughts. Victoria started to talk about her long 15 hours' working days and said that she was never home at the same time as Alfonso. While speaking about her workload she said: "I am asking that I am the most important thing in his life ... I don't want to be in a relationship where I am not the first one" (t74). She gave some examples of what she meant: "Some text message, or some small surprise ... or maybe that he has hovered ... (t87)" Alfonso replied that he was committed but that Victoria did not recognize it: "I feel that I do things for us and for our relationship. I don't understand how you don't see that." (t99) When the male therapist asked about the assignment that he had given to them at the end of the previous session, they both said that, as such, it had been a good task. For the most part of the end of the session, they wanted to speak of Alfonso's upcoming trip to his home country.

At a certain moment, Alfonso referred to an issue he had with his parents: "there's some kind of, some kind of problem that I have with them, so maybe if I could maybe solve that kind of thing then maybe I could also, this would be also easier that you [to V] could come there, but I feel (...) it's that's just the way I feel about them somehow, I just feel too forced to, now to just change that, it feels something not natural ..." (t409) He explained that when he visits his family, his mother would want him to be with her all the time, and that probably when Victoria was around she felt jealous of her (t499). By the end of the session, Victoria and Alfonso seemed to agree that for the time being their idea of Victoria not visiting Alfonso's family was a good solution (t513).

At the outset of **the third session**, Victoria said that she has started working less (t14), and that they had more time to see each other (t16). It was "a good period" (t24), Alfonso added. Victoria and Alfonso recounted that during Alfonso's visit to his home country, he maintained regular contact with Victoria. She experienced this increased sensitivity to her as an expression of a shift in his loyalty and commitment to her. There had been a fight about Alfonso going to a concert in a neighboring country before he went to his home country to visit his family, without talking to Victoria about it (t37). But in the end it was no big deal to Victoria, as when Alfonso came back she could see that he had missed her (t71). After the therapist asked them how they wanted to use the time (t81), they reassured the therapist that things were fine (t81) but then returned to their misunderstandings and their fighting. Both agreed that it often started with Alfonso's reaction to a simple question of Victoria's. According to Victoria, it is a very big problem for them (t129): "He doesn't trust me at all," Victoria stated (t131). Alfonso explained that his problematic reaction to a question of Victoria's happened when he had to explain something to her or prove something (t143). For Victoria, it sounded as if she was not allowed to talk about things and as if she did not have the right to feel sad (t148). Alfonso emphasized that it was important for him that Victoria would understand that he couldn't control his reaction to her question: "I ...can't control it, and ... I ... don't want to have this kind of reaction, but it is just how it is, I don't control it...." (t164) The therapists explored a possible link of Alfonso's reactions with his childhood and his experience of being hit by his mother (t262). The cultural difference between a Scandinavian and a Mediterranean culture was discussed in terms of emotional reactions (t278). Then Victoria explained that when Alfonso reacted in this problematic way, she started crying (t310). After this overview of exploring their circle of interactions, both partners agreed that they now had found a better way to deal with their troublesome communication: "At first we tried to talk about it, but then we realized that it's not any use," Victoria explained (t346). They have learned that it doesn't do any good to try to solve it by talking (t348), and they decided to take some time and keep some distance instead and let things settle on their own: "... when we see that it's like that, we don't talk to each other and we left things calm down... before we were just trying over and over, to solve the thing, that was just worse..." Alfonso explained (t354). He added that when they come back together after such a tense situation they start hugging (t376). Asked by the therapist who takes the initiative, Alfonso said: "I think it is always me" (t378). In a final reflection together the therapists highlighted that Victoria and Alfonso seemed to have found new ways of being together that are more constructive (t405).

The fourth session was brief. It started with the therapist asking "How is your life today?" (t1). Victoria replied "It's fine" (t2) and Alfonso agreed (t3). They both said it had been a good period (t12–13) because they didn't have their usual fights (t15). "How did you make it possible ... not having fights?" the therapist inquired (t26). "We have more time together," Victoria first replied (t28), but on second thought they didn't know how they had done it. The therapist inquired if Victoria changed her way of questioning Alfonso (t49), and Alfonso said it was the same (t52), but that there were fewer fights (t59). Victoria then said that she had been able to take things less seriously (t70): "...for example, if I get this bad feeling nowadays it's easier for me to let go..." (t74). Alfonso backed her up by saying that he had noticed that (t75). Victoria explained that she had been thinking a lot and that she realized that she had these issues with trust and now she forces herself to believe Alfonso's words of commitment in spite of the inner voice that says "don't believe anything you hear" (t83). She emphasized several times that this is a struggle for her: "I know I will never be completely normal" she said (t93). Referring to the fact that there had been no fights in the last weeks the therapist addressed the couple and asked: "...is it the right or the wrong conclusion that you have ...learned...to do

something different, together?" (t111–113). They both agreed, but Victoria emphasized that it is never easy to learn (t117). She further said that they had learned to
listen to each other (t121), and Alfonso added that they had learned to deal with the
threat of an escalation by saying "OK, let's just now stop" (t122). The therapist then
asked about possible challenges for them in the future. One challenge they mentioned is that they are going to move in 2 days (t141–160). Another challenge would
be Alfonso's next trip to his home country, which was scheduled for 5 weeks later
in the Christmas season. Alfonso concluded that he was confident that—like his
previous trip a month earlier—all would go smoothly (t162). Victoria replied that
for her it was weird to spend Christmas in different places, but that she was not
ready yet to go with Alfonso to his home country (t169). Then they turned again to
talk about moving and about the different views they have concerning their home.
Rounding up the session, the therapist then asked if the couple now had an answer
to the question they came in with (t235). "Yeah, I think we are getting better, yeah,"
Victoria replied (t238). "Yes," Alfonso agreed (t239). They decided to end the session there and not to have any more therapy sessions.

References

Duncan, B. L., Miller, S. D., Sparks, J. A., Claud, D. A., Reynolds, L. R., Brown, J., et al. (2003).
    The Session Rating Scale: Preliminary psychometric properties of 'working' alliance measure.
    Journal of Brief Therapy, 3(1), 3–12.
Miller, S. D., Duncan, B. L., Brown, J., Sparks, J. A., & Claud, D. A. (2003). The Outcome Rating
    Scale: A preliminary study of the reliability, validity, and feasibility of a brief visual analog
    measure. *Journal of Brief Therapy, 2*(2), 91–100.

Chapter 3
The Development of Dialogical Space in a Couple Therapy Session

Peter Rober

In this chapter, we want to present a method to research the development of dialogical space in a couple session. It is an exploratory, retrospective, microanalytic research method. Like the other research methods that are presented in this volume, it is a qualitative method, which means that it is not focussed on a theory driven hypothesis testing. Rather, it is meant to help the researcher notice things that would otherwise remain unnoticed, in order to arrive at a better and more nuanced elucidation of the process of therapy.

In this chapter, we will first sketch the theoretical frame of the method. Then we will discuss the three stages of the method and illustrate the use of the method by applying it to the first 35 turns of the first session of Alfonso and Victoria's therapy. Finally, we will discuss some of the most important observations and our enriched understanding of the dynamics of the dialogue between Alfonso, Victoria, and the therapist.

Dialogue and Dialogical Space

Especially inspired by the Russian thinker Mikhail Bakhtin (1981, 1984, 1986), the concept of dialogue became very important in the field as an answer to the ethical challenges of family therapy practice (e.g. Rober, 2005; Seikkula & Olson, 2003). It is interesting, however, that Bakhtin used the concept dialogue in two distinct ways: as a prescriptive concept and as a descriptive concept (Stewart, Zediker, & Black, 2004). When dialogue is used as a prescriptive concept, the term is reserved to refer to a particular kind of interaction of a high quality. Dialogue then is the

P. Rober (✉)
Department of Neurosciences, Institute for Family and Sexuality Studies,
KU Leuven and Context, UPC KU Leuven, Belgium
e-mail: peter.rober@med.kuleuven.be

© Springer International Publishing Switzerland 2016
M. Borcsa, P. Rober (eds.), *Research Perspectives in Couple Therapy*,
DOI 10.1007/978-3-319-23306-2_3

opposite of monologue. In Buber's view, for instance, dialogue (I–Thou) is ideal and monologue (I–It) should be avoided. As Stewart et al. (2004) write, prescriptive approaches make ethics central.

In the marital and family therapy (MFT) literature, the concept of dialogue is often used in such a prescriptive way highlighting the ethical ideal. Usually dialogue is seen as the opposite of monologue, implicitly suggesting that good therapy is dialogical, while bad therapy is monological or arguing that clients enter therapy with fixed, monological stories and that therapy consists of dialogising these stories (e.g. Penn & Frankfurt, 1994). In that way, dialogue implicitly is described as an ideal endpoint of a process moving from monologue to dialogue.

The concept of dialogue in Bakhtin's work, however, is complex (Vice, 1997) and simply describing it as the opposite of monologue does not do justice to the wealth of his work. Indeed, dialogue is described by Bakhtin, not only as a prescriptive concept, but first and foremost it is presented as a descriptive concept. In that way, the concept focusses on epistemological issues and it highlights the relational and interactional character of all human meaning making: All language is dialogic. In this perspective, monologue can also be understood as a part of dialogism and we can speak of dialogical dialogues and monological dialogues (Morson & Emerson, 1990).

In the context of this descriptive view of dialogue, Stewart et al. (2004) highlight the importance of tensionality in Bakhtin's work. According to Bakhtin, in an ongoing conversation there is a continuous dynamic tension between the monological and the dialogical functions, of which Bakhtin scholar Caryl Emerson writes: "Dialogue is by no means a safe or secure relation. Yes, a 'thou' is always potentially there, but it is exceptionally fragile; the 'I' must create it (and be created by it) in a simultaneously mutual gesture, over and over again, and it comes with no special authority or promise of constancy. … Imbalance is the norm". (Emerson, 1997, pp. 229–230). According to Bakhtin, life is an ongoing, unfinalisable dialogue continually taking place (Morson & Emerson, 1990). Bakhtin (1981) does not characterise dialogue as something peaceful or at rest, but rather calls dialogic life "agitated and cacophonous" (p. 344). What is said in dialogue is the product of dynamic, tension-filled processes in which two tendencies are involved: the centripetal (centralising and unifying) forces and centrifugal (decentralising and differentiating) forces (Bakhtin, 1981; Baxter, 2004; Baxter & Montgomery, 1996). The *centripetal* stands for a structured order dialogue strives for. This could be a single story, an agreed upon explanation, an accepted solution, a contract, homogenity, harmony, etc. The order comes at the expense of things left unsaid, facts overlooked, experiences not noticed, words remaining unarticulated, etc. In contrast, the *centrifugal* stands for the disruption of the order and the messiness of things, unforeseen complexities, heterogenity, conflict, the scattered details that are unexplained and that unsettle the account, and so on. In dialogue, these opposing forces are in constant dialectical tension; one being the antithesis of the other. Contrary to Hegelian dialectics that prescribes the finalisation of dialectic tensions in a synthesis, according to Bakhtin these dialogical processes are unfinalisable: the tension between the two opposing forces never reaches a final solution.

The tension between centripetal and centrifugal forces finds expression in what is actually said in a conversation, and in what is not said. At any one moment in a dialogue some things have been said; other things may be said later; and still other things will never be said. This simple observation is very important as it has far reaching consequences for a therapist, as well as for a researcher. Theoretically, it can be connected with several concepts that characterise dialogue as a never-ending, interpersonal process. These concepts refer to each other in multifaceted ways.

- Dialogue is a process in time. There is always *before* and *after*. In narrative psychology the concept of <u>sequentiality</u> is used (Bruner, 1990). An utterance is not an isolated message or expression of the individual speaker with his/her inner motivations and intentions. An utterance has meaning in a context of time and place (chronotope, cfr. Bakhtin, 1981). Whatever is said becomes meaningful by the place it occupies in a sequence of events (Linell, 1998; Markova, 2003).
- <u>Subjectivity</u>: This is about the centre of experience that each of us is. Our subjectivity is largely internal conversation between inner voices within ourselves. This inner conversation comes into being through the continual dialogical process with others (Linell, 2009). Inner conversation accompanies outer conversation, in the sense that what we say only partially reflects what we are thinking. Part of our thinking remains private and unarticulated, sublingual, and inchoate (Lewis, 2002).
- Here the principle of <u>selectivity</u> comes in: Some things will be shared with others, other things won't. Besides this selectivity in content, there is also a selectivity in timing: some things are said earlier, other things are said later. Some things are never said. This selectivity does not come out of the blue, but it is part of the context. In other words, this selectivity is responsive (Linell, 2009): we respond to some things (while we ignore other things) and by responding to them we validate them as important (retro-construction). Of course this also connects with the concept of sequentiality (see above).
- <u>Responsivity</u>: Utterances in dialogue are other-oriented. Whatever is said is always said in response to what has been said before (Linell, 2009). Also, everything that is said is an invitation to the others to respond. In that way the participants shape the dialogue together. This also connects with the concept of selectivity (see above). As Linell (2009) writes, "Every act is selectively responsive…" (p. 167) in the sense that we do not respond to everything, but that there is a selection in our responses: to some things we respond, while other things we neglect.

Based on the concepts of sequentiality, subjectivity, selectivity, and responsiveness, we can now define the concept of dialogical space. Dialogical space refers to the virtual environment of expectations and entitlements about what can be talked about in a certain chronotope. In other words, it refers to what is said at any given moment in the conversation and implicitly it also refers to what is not (yet) said. It is a concept that rests on the assumption that it would be possible to freeze a moment in time in the dialogical process: whatever is said up until that moment is part of the dialogical space. Of course, this is an abstraction as in a dialogue there is never a moment of standstill. Dialogue is never a tranquil state of things. Rather, it is a restless process through time, in which there is a constant tension between what is said and what is not said. Dialogical space, then, refers to what is acceptable to the participants to discuss at a certain moment in the dialogue. It refers to the room to talk about what obviously—usually without overt negotiation—fits the shared implicit agenda of the participants.

Researching Dialogical Space

In this chapter, we want to present a method to research dialogical space in the context of a couple therapy. We will present this research method against the theoretical background of three traditions: (1) Goffman's dramaturgical theory of human interaction (Goffman, 1959), (2) conversation analytic focus on sequential organisation of talk (Vehviläinen, Perälylä, Antaki, & Leudar, 2008), and (3) theory of the dialogical triad (Markova, 2003).

1. Goffman (1959) described human interactions as theatrical performances. Although his theory is very individually oriented as the self of the person is central, his emphasis on the *performative* aspect of human interaction and especially his distinction between the *on-stage area* and *the backstage area* are interesting for our analysis. This distinction connects with the distinction made by Anderson and Goolishian (1988) between the said and the not-yet-said, and their description of marital and family therapy as a process of gradually making room for what has not been said yet. In Goffman's terms, therapy could be described as a performance in which gradually more things from back stage are presented on stage. This is the process we are interested in, and with our research method we want to be better able to describe this process in a multiactor psychotherapeutic setting.

2. The basic conversational analytic strategy of *taking what people are doing, saying, not-saying, at a particular moment, in a particular manner, and trying to find out the kind of problem which it might be a solution for* (Ten Have, 1999), is also the basic strategy of our way to study the development of dialogical space. Furthermore, CA's focus on the *sequential organisation* (Schegloff, 2007) is of interest to us. The significance of utterances in an interaction depends on their position in a sequence of utterances (Linell, 1998). In an interaction sequence some actions call for a response. Other actions are such responses. Therefore, Vehviläinen et al. (2008) make a distinction between *initiatory* and *responsive* actions. However, in a dialogical view, any initiatory action is a responsive action, and any response invites a response in its turn. Still, as we will see, this distinction between initiatory and responsive actions can be useful in the study of the development of dialogical space, in that sometimes we can see that someone introduces a theme for the first time. Such an introduction can be seen as an initiative that serves as an invitation to the participants of the dialogue to talk about the theme.

3. Another theoretical background for our research method is Markova's dialogical triad: Ego-Alter-Object (Markova, 2003): a person (Ego) talks to another person (Alter) about something (Object). For instance, in a specific utterance the Ego and the Object can be more or less identical; for instance, when a person talks about himself. So when a woman talks to her husband and says "I'm sad", in addressing her husband (Alter) the woman (Ego) talks about her own emotions (Object). Furthermore, it is clear that for Markova, the relationship between the three components of the triad is a dynamic one. *Dialogical tension* is the key word in her model (Markova, 2003). According to Markova, tension is implicit in all dialogical situations, as it is the source of change. It is the presence of dialogical tension that makes the triad Ego-Alter-Object dynamic, rather than static (Markova, 2003). Another complexity is the Alter. This concept refers to the Bakhtinian concept *addressee*, and in multiactor conversations it refers to *multiple addressees*. This is the challenge for all dialogical research in family therapy and also for our research on the dialogical space.

By definition dialogical space is difficult to investigate because of its focus on the dynamic of what is said and what is not said. Rogers et al. (1999) formulate the challenge as follows: "If we assume, as we do, that the unsaid can contribute something valuable to our understanding of how an individual understands the world, then what language can we use to present what is unsaid?" (p. 80) Indeed, how do we know what is not said? In principle we do not, as what is not said remains hidden in the person's secret garden. In the context of studying dialogical space in a therapy session, our answer to this question is the following one: we can look at the development of what is talked about in the session.

Our retrospective, microanalytic method, while meant to research marital or family therapeutic sessions, was originally inspired by dialogical research methods used in the context of individual therapy, like the Dialogical Sequence Analysis (Leiman, 2004) and Positioning Microanalysis (Salgado, Cunha, & Bento, 2013). These methods are based on the dialogical triad Ego-Alter-Object (Markova, 2003). As we will later explain in more detail, in our research method, we start from one component of Markova's dialogical triad: the referential Object, or what Wortham (2001) would call *the narrated event*. We could also call it the theme of the conversation. The two other components in the first stages of the method are the background of the evolution in the referential object throughout the dialogue. In the last stage, they become more central.

We will now describe the research method in more detail.

Research Method

Our research method is a retrospective, microanalytic method focussed on the development of the dialogical context in a marital or family therapeutic session. Based on the three theoretical sources of inspiration we discussed above, we can further operationalise the focus of our research method. In this method the time dimension is central, as the basic question of this method is: *what is talked about when?* Time in this context does not refer to actual time, but rather to sequential time, in which one utterance comes before the next.

The starting point for our research method is a detailed transcript of a MFT session and if possible the videotape of the session. Our research method has three stages:

Stage 1. Retrospectively listing the themes that have been discussed: We look at the whole transcript and we list the themes that were discussed in the session. Here we focus on the content of the client's story in the session; on what can be also called *the narrated event* (Wortham, 2001), or *the referential Object* (Markova, 2003).

Stage 2. Tracking the sequential organisation: We summarize the client's story and then track what is told first, and what then, and what then, … Here we introduce the time dimension and we focus on the topical history of the dialogue.

Stage 3. The initiatives and responses: Now we take a conversation analytic stance and try to understand for each topical sequence what created the dialogical space in which it was told. This means that we look at initiatives to address a new theme and at replies to these initiatives. A response to an initiative can be or accepting, or declining. Considering Markova's dialogical triad, we can say that in this third stage the emphasis shifts from the Object to what happens between the Ego and the Alters.

Through these three research stages we try to understand how the dialogical space evolves throughout the session. This means focussing on what is talked about and what is not yet talked about, with a special attention to initiatives to address a new topic, and to the ways in which such initiatives are responded to by the participants.

The Case of Alfonso and Victoria[1]

In this chapter, we will use our research method to look at the case of Alfonso and Victoria. We will illustrate its use and potential by focussing on the first 35 turns of the transcript of the couple therapy session of Alfonso and Victoria (for an overview of the therapy, see Chap. 2). In the first 35 turns, the initial problem story is presented with minimal intervention from the therapist. In turn 36 the therapist, for the first time, does not follow the lead of the couple but explicitly addresses Alfonso and decides on taking a focus (Alfonso's fear).

---

[1] In this section the numbers between brackets refer to the turn in the conversation.

Stage 1. The Story

In order to get the story straight, the transcript is read several times. In the margin key words or brief sentences are written down. These are words that seem to catch the referential object of the dialogue. In the choice of words it is important to stay as close as possible to the words the participants of the conversation actually have used. We bring it all together in a table to get some overview. In the first column we list the referential objects, in the second column we refer to the turns and to the person speaking (Alfonso, Victoria, or T), and the third column is a column for memos (Table 3.1).

Stage 2: The Sequential Organisation

Then a list is made of all the referential objects. In the case of Victoria and Alfonso this was the list for the first 35 turns: the Outcome Rating Scale (ORS), depressed, better now, scars in the relationship, not as much patience in listening, I just can't, panic, afraid that it might be the same, not the same strength and patience, getting better, we don't talk anymore, I don't have the right to feel bad. Looking at the themes, we could summarize the story as *I used to be depressed, now I'm better, but now we have a relational problem.*

Table 3.1 The story as it develops

Story as it develops	Turn nr.	Comments
Filling out the ORS form	t1-7	
Victoria got depressed	t8V	
This resulted in "scars in the relationship"	t10V	V also suggests that she got depressed because she found it difficult "to start to trust someone, to feel loved, and to feel love"
A does not have as much patience in listening as he used to have	t14A-27A	
There is a suggestion of "irritation" (t20A). This suggestion is picked up by the therapist (t21T). But then V takes over and talks about A's fear ("afraid") (t22V)	t20-21-22	
He is afraid that it could become "some similar situation"	t22V-29A	
V talks about A's "panic"	t28V	
This results in "we can't talk anymore" and the feeling that I don't any more have "the right to feel bad"	t34V	V talks about a lack of patience until now. Only in t29, he talks about being "afraid". He refers to "a similar situation". He does not openly talk about depression

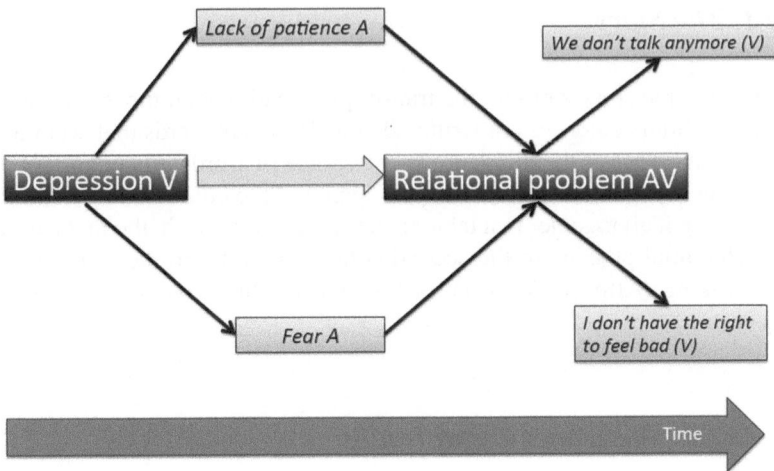

Fig. 3.1 The sequential organisation

Bringing in the time dimension, the story of the first 35 turns in the sequential organisation could be summarized like this (Fig. 3.1).

Stage 3: Initiatives and Responses

While in the first and the second stage of the research the focus is on the referential object, now we concentrate on the two other components of the dialogical triad: the dialogue between the participants. We look at the details of how the themes emerged, who took the initiative, how the participants responded, and how the themes developed further during the session. Here we want to focus on four observations:

1. Victoria takes the initiative to start to talk about what she calls *the beginning* (t8)
2. Victoria takes the initiative of suggesting the possible cause of her depression (t10)
3. Alfonso's initiative to focus on *now* and on his responsibility in what is difficult in their relationship (t14–22)
4. Alfonso's hesitating initiative to talk about irritation was not picked up (t20–22)

Let's focus on these four observations and consider them in more detail:

1. V takes the initiative to start to talk about what she calls *the beginning*: her depression (t8). Of course, calling the depression the beginning is remarkable on its own, especially since we know from the rest of the therapy that the depression only started 2 years before and they had been living together for 3 years. So a lot of things must have happened in the relationship before the so-called *beginning*. This highlights that starting the story from the depression and deciding that is the

beginning is less obvious than it sounds. So the question can be posed how we can understand that Victoria choose the depression as the beginning of the story? Let us consider some of the possible answers to this question. The theme of the depression may have been primed by filling out the ORS forms. The ORS focusses on the outcome of therapy and maybe this reminded her of her individual therapy when she was depressed. Another way to understand the choice of the depression as beginning is that Victoria's depression could be seen as reflecting the couple's legitimation of their choice for therapy. It is accepted in our society that you seek therapy because of a psychiatric condition. This can help us understand that something that can count as an *official* problem, the depression, was invited front stage at the outset of the therapy and was *the beginning*, as Victoria calls it. In that way, possibly the depression was a legitimation of the therapy. This legitimation then may have served as a reassurance for the couple and for the therapist in the sense that it reminded all parties: *we are in the right place; it is legitimate for us to be here together.* In t10 she explicitly refers to "the reason why we are here", suggesting that this has been on her mind in the beginning of the session. By taking her depression as a starting point she also puts the focus on herself. According to the story she tells in the session, she started it all. Furthermore, depression is described as an individual condition. In that way in a sense Victoria takes responsibility for her condition. It is remarkable that she is the only one who calls it depression (t10 "I got depressed").

2. Victoria's initiative of suggesting the possible cause of her depression: according to her story she got depressed because she found it difficult "to start to trust someone, to feel loved and to feel love" (t10). While the issue of trust in the relationship is introduced here, but it was not developed further at this moment. This is interesting, especially because it reappears later in the session (t24, t34), but also in the next sessions. It will even become one of the main themes of the therapy. The question therefore can be posed, if this theme is so important to become one of the main themes of the therapy, why it was not further discussed in the beginning of the first session? When we look closer at the responses of Alfonso and the therapist, we notice that neither of them picks up the theme. The therapist responds to Victoria in the next turn (t11), but only to refer to Alfonso's name. Also Victoria herself did not pursue it further in the next turns: she only mentions it in t10, but then she lets it go. She does not insist on talking about it further. This is interesting and it begs the question: how can we understand this? We cannot be sure how to answer this question, but the way the dialogue develops suggests that this issue of trusting someone is sensitive for Victoria. It seems that she feels vulnerable about it. Could it be that she judged this theme to be too sensitive as to address so early in the therapy? Or was it because her initiative was not responded to with more support from Alfonso and the therapist?

3. Alfonso's initiative to focus on *now* and on his responsibility in what is difficult in their relationship. Rather than acknowledging that Victoria is responsible for her depression or pursuing the theme of trust in the relationship, Alfonso responds by referring to the situation now (t14 "We have this kind of situations"). From his way of talking (e.g. t14; starting sentences he does not finish, being

very confusing and hard to understand, …) we get the general impression that Alfonso is nervous and very cautious in his choice of words. Then finally, at the end of t14, he finishes a sentence and he takes responsibility for the situation: referring to the past he says *"…I had more patience in listening"*. This is remarkable because without mentioning the depression, he refers to the period of the depression, in order to characterise the situation now and making sure that he is taking responsibility for their difficulties.

In response to Alfonso's initiative to talk about *now*, and on his responsibility in what is difficult in the relation, Victoria is very accepting. She backs Alfonso up in the telling of his story, by suggesting words (t15) and by offering clarifying examples (t17). This support of Victoria shows that dialogical space at that moment was opened to talk about Alfonso's contribution to the relational problems.

4. Alfonso's hesitating initiative to talk about irritation was not picked up as a theme to talk about (t20, 21, 22): With a lot of hesitation, Alfonso seems to want to say something about irritation (t20), at least that is the way the therapist interpreted Alfonso's hesitation (t21), and then Victoria intervenes and takes the conversation away from the theme of "irritation" and steers the conversation in the direction of fear (t22: "…he's afraid…"). This can be considered a sequence in which Alfonso hesitatingly takes the initiative to open space to discuss his irritation. The therapist helps him to express his irritation, but then Victoria rejects it as a theme for discussion as she starts to talk about fear. In that way she closes the dialogical space to talk about irritation, by proposing to talk about Alfonso's fear. This invitation of Victoria to talk about fear, rather than about irritation, is accepted by Alfonso and also by the therapist. Neither of them takes the initiative to try to again pick up the theme of irritation. They seem to go along with Victoria and prefer to talk about fear at this point in the conversation.

Discussion

As with a lot of qualitative research methods our method is exploratory. It is not focussed on the testing of hypotheses derived from theories, but it is focussed in the first place on helping us researchers to notice things that are remarkable and to carefully describe these things. That is why we have focussed in the previous section on observations. These observations will be the starting point of the discussion of our research findings.

In our analysis of the first 35 turns of the couple therapy of Victoria and Alfonso, we should not make the mistake of only focussing on what is actually said in these 35 turns. It is of the essence in this research approach to take as a starting point the whole of the client's story that has developed in the four sessions, and then to look at what was not said in the sequences under consideration (in this research, the first 35 turns). When we consider the whole therapy (the four sessions), arguably the issues of Alfonso's family, their cultural differences and of Victoria's difficulty to trust someone could be pointed to as the most important themes of the therapy as they seem to be at the basis of their conflicts. It is remarkable that these themes are not addressed in the first 35 turns of the first session. So when the therapist asks "Where would you start?" (t7), the clients did not start with what would later prove to be the most important in their therapy. Instead of talking about trust or about Alfonso's family, Victoria chooses to start from what she calls the beginning (her depression), rather than from these issues that, in the context of the whole couple therapy will prove to be more central, more emotionally charged, and more conflict prone.

In these first turns, Victoria generally takes the initiative. For instance, she chooses to start to talk about the beginning, and she decides that the beginning is her depression. Alfonso seems to be more reluctant to talk. He lets Victoria take the initiative, and especially in the beginning of the session what he says is hazy and he is cautious in his choice of words, starting sentences he does not finish, restarting another sentence, looking for the right words, etc. However, Alfonso takes initiatives too. He does this in a cautious way, making sure not to put Victoria on the spot. For instance he steers away from the subject of Victoria's depression, focussing on the current difficulties in the relationship and taking the blame for them. Of course, by taking the blame he makes sure that she doesn't feel blamed. Indeed, as often is the case in couple conflicts, blaming may be central in their discussions at home. It would make sense that they are very careful to try to avoid blaming in order to make the therapy work, and avoid that it might run aground in the kind of hopeless conflicts as they have at home (we know this from the further therapy). This can help us understand too why the topic of irritation was not accepted as a topic of discussion by Victoria, followed by an implicit agreement of Alfonso and the therapist. It might be difficult to talk about irritation without implicitly or explicitly blaming someone (the person one is irritated about).

All in all, it gives the impression that both Alfonso and Victoria are very cautious about what themes to address in the beginning of the sessions. It seems as if at this stage talking about individual emotions (fear) and experiences (lack of patience) is preferred over talking about relational things, probably because of the risk of conflict and blaming. But already after a few turns, they are framing the reason to go to therapy in a relational way (e.g. *we don't talk anymore...*). So in general, there seems to be a move in the first 35 turns from individual framings towards more relational framings.

To recapitulate, we have conceived the development of a dialogical space as a process comprised of initiatives and responses, leading to the opening or closing of space to address certain topics. We can summarise our observations as follows:

- *When dialogical space is opened* for a certain topic (for instance, the topic of Victoria's depression in t8), participants implicitly agree that it is acceptable to talk about this topic at that particular moment in that dialogical context. Based on our observations we can hypothesise that often the opening of dialogical space starts when one of the participants takes the initiative to coin a phrase referring to a new topic and invites the other participants to respond. It would be conceivable that such an invitation to talk about a certain topic would be an explicit question (e.g. "shall we start with talking about my depression?"). In the very first turns of the first session of Alfonso and Victoria, however, the invitations were more implicit, as were the responses. The response of the other participants to such an invitation may open the space (e.g. when the invitation is not challenged) or close it again (when the invitation is refused, neglected, …). Our research of Alfonso and Victoria's session suggests that opening a dialogue to new topics may involve a risk (e.g. the risk of conflict, blame, feeling guilty, …). Introducing a new topic can be seen as the disruption of a kind of equilibrium or an established order of the dialogue: a threat of a kind of *pax dialogica* for the participants. In Bakhtin's language the opening of dialogical space for a new topic is often part of the *centrifugal* tendency of a dialogue.

- When *dialogical space is closed* for a new topic, in Bakhtin's language, this is the centripetal tendency of dialogue, striving for order, certainty, and repetition. In our research we could observe that talking about Alfonso's irritation was avoided, in implicit agreement between the three participants. There was a cautious invitation, but it was not accepted by Victoria and then Alfonso and the therapist went along with the dialogical path Victoria proposed.

Limitations and Suggestions for Further Research

The research we reported on here is—as far as we know—the first research in its kind. Like all types of research it has limitations. The main limitation in our opinion is that the input of the clients is not incorporated in the research. It would have been interesting to know how Alfonso and Victoria looked back at the first 35 turns of the first session and how they understood the development of the dialogical space in retrospect. A tape-assisted recall procedure would have been useful in order to help them to recall how they experienced this section of the first session (Kagan, 1975; Rober, Elliott, Buysse, Loots & De Corte, 2008a, 2008b). However, as in this project the data set consisted of the transcripts of the four sessions, and nothing else, we did not have access to the clients' experiences.

Furthermore, it is important to acknowledge that this is the first study of this kind and that it would of course make sense to use the method presented here to study other couple therapies. Also, it would be interesting to use this method to study family therapy sessions.

Besides the mere replication of this research, these are some avenues that seem to us to be, not only very interesting, but even necessary to further understand the development of dialogical space in a MFT session:

- As we wrote in the section on theory, the concept of selectivity suggests that topics might remain unspoken for some time, until the time seems ripe to discuss them. It is very important to study what exactly are the implicit or explicit considerations of clients to open space for an important topic? In order to study this, one could work with data on inner conversations of dialogical participants. For such a research, tape-assisted recall methods could be used (e.g. Rober et al., 2008a, 2008b).

- Also, it would be interesting to study carefully what could be the therapist's contribution to the opening of dialogical space for the discussion of important themes. From other research (e.g. Rober, Van Eesbeek, & Elliott, 2006) we know that a therapist can not only contribute to the opening of space, but also to the closing of space. Furthermore, from this research we also learned that therapist can open or close dialogical space without being aware of it. This means that we would need an observational study design to better understand the therapist's contribution to the development of dialogical space, rather than a design based on self-report.

Conclusion

In this chapter, the concept "dialogical space" refers to the virtual environment of expectations and entitlements about what can be talked about. Our study of the development of dialogical space of this couple therapy session illustrates that dialogical space might be a useful concept in understanding implicit dynamics in couples. Our research suggests that closing dialogical space may create some stability in an uncertain and tension-filled dialogue, as it constrains what can be talked about in the present moment of the given conversation. Our findings seem to indicate that closing space for sensitive issues in couple therapy can help the participants at the outset to keep the *pax dialogica*, and avoid conflicts and escalations that might endanger the budding therapy. The closing of dialogical space results in topics that remain—at least for the moment—unspoken in the session. As happens later in the therapy of Victoria and Alfonso, space to discuss these themes may be opened later and in this the help of the therapist can be useful. Looking at the development of the dialogical space offers a perspective on couple therapy in which the continuous tension between invitations and responses to these invitations result in the dialogue that has a sense and a direction, and that develops and enriches through time.

References

Anderson, H., & Goolishian, H. (1988). Human systems as linguistic systems. *Family Process, 27*, 371–393.

Bakhtin, M. (1981). *The dialogic imagination*. Austin, TX: University of Texas Press.

Bakhtin, M. (1984). *Problems of Dostoevsky's poetics*. Minneapolis, MN: University of Minneapolis Press.

Bakhtin, M. (1986). *Speech genres & other late essays*. Austin, TX: University of Texas Press.

Baxter, L. A. (2004). Dialogues of relating. In R. Anderson, L. A. Baxter, & K. N. Cissna (Eds.), *Dialogues: Theorizing difference in communication studies* (pp. 107–124). London: Sage.

Baxter, L. A., & Montgomery, B. M. (1996). *Relating: Dialogues & dialectics*. New York: Guilford Press.

Bruner, J. (1990). *Acts of meaning*. Cambridge, MA: Harvard University Press.

Emerson, C. (1997). *The first hundred years of Mikhail Bakhtin*. Princeton, NJ: Princeton University Press.

Goffman, E. (1959). *The presentation of self in everyday life*. New York: Doubleday Anchor.

Kagan, N. (1975). *Interpersonal process recall: A method of influencing human interaction*. Houston, TX: Educational Psychology Department, University of Houston-University Park.

Leiman, M. (2004). Dialogical sequence analysis. In H. Hermans & G. Dimaggio (Eds.), *The dialogical self in psychotherapy* (pp. 255–270). Hove: Brunner-Routledge.

Lewis, M. (2002). The dialogical brain: Contributions of emotional neurobiology to understanding the dialogical self. *Theory and Psychology, 12*, 175–190.

Linell, P. (1998). *Approaching dialogue: Talk, interaction and contexts in dialogical perspectives*. Amsterdam: John Benjamins.

Linell, P. (2009). *Rethinking language, mind, and world dialogically: Interactional and contextual theories of human sense-making*. Charlotte, NC: Information Age.

Markova, I. (2003). *Dialogicality and social representations: The dynamics of mind*. Cambridge: Cambridge University Press.

Morson, G. L., & Emerson, C. (1990). *Mikhail Bakhtin: Creation of a prosaics*. Stanford, CA: Stanford University Press.

Penn, P., & Frankfurt, M. (1994). Creating a participant text: Writing, multiple voices, narrative multiplicity. *Family Process, 33*, 217–231.

Rober, P. (2005). Family therapy as a dialogue of living persons: A perspective inspired by Bakhtin, Volosinov & Shotter. *Journal of Marital & Family Therapy, 31*, 385–397.

Rober, P., Elliott, R., Buysse, A., Loots, G., & De Corte, K. (2008a). What's on the therapist's mind? A grounded theory analysis of therapist reflections. *Psychotherapy Research, 18*, 48–57.

Rober, P., Elliott, R., Buysse, A., Loots, G., & De Corte, K. (2008b). Positioning in the therapist's inner conversation: A dialogical model based on a grounded theory analysis of therapist reflections. *Journal of Marital and Family Therapy, 34*, 406–421.

Rober, P., Van Eesbeek, D., & Elliott, R. (2006). Talking about violence: A micro-analysis of narrative processes in a family therapy session. *Journal of Marital and Family Therapy, 32*, 313–328.

Rogers, A. G., Casey, M. E., Ekert, J., Holland, J., Nakkula, V., & Sheinberg, N. (1999). An interpretive poetics of languages of the unsayable. In R. Josselson & A. Lieblich (Eds.), *Making meaning of narratives (The narrative meaning of narratives)* (Vol. 6, pp. 77–106). London: Sage.

Salgado, J., Cunha, C., & Bento, T. (2013). Positioning microanalysis: Studying the self through the exploration of dialogical processes. *Integrative Psychological and Behavioral Science, 47*(3), 325–353.

Schegloff, E. A. (2007). *Sequence organisation in interaction: A primer in conversational analysis*. Cambridge: Cambridge University Press.

Seikkula, J., & Olson, M. E. (2003). The open dialogue approach to acute psychosis: Its poetics and micropolitics. *Family Process, 42*, 403–418.

Stewart, J., Zediker, K. E., & Black, L. (2004). Relationships among philosophies of dialogue. In R. Anderson, L. A. Baxter, & K. N. Cissna (Eds.), *Dialogue: Theorizing differences in communication studies* (pp. 21–38). London: Sage.

Ten Have, P. (1999). *Doing conversational analysis: A practical guide*. London: Sage.

Vehviläinen, A., Perälylä, A., Antaki, C., & Leudar, I. (2008). A review of conversational practices in psychotherapy. In A. Perälylä, C. Antaki, A. Vehviläinen, & I. Leudar (Eds.), *Conversation analysis and psychotherapy* (pp. 188–197). Cambridge: Cambridge University Press.

Vice, S. (1997). *Introducing Bakhtin*. Manchester, UK: Manchester University Press.

Wortham, A. (2001). *Narratives in action: A strategy for research and analysis*. New York: Teachers College Press.

Chapter 4
Introducing Novelties into Therapeutic Dialogue: The Importance of Minor Shifts of the Therapist

Aarno Laitila

In this chapter, I paint a picture of the evolution of the therapist's actions in the beginning phases (during the first 70–102 speech turns depending on the session) of four consecutive couple therapy sessions of a young intercultural couple (see Chap. 2). The outcome of these therapist actions is considered both in the microanalytical context of the immediate conversation negotiating the couple's relationship as an in-session event and also in the context of this four-session process, suggesting even the long-term effects of the treatment.

The beginning of each psychotherapy process embodies condensed, vital information that will concern the entire therapy process. Some psychotherapists have made claims that the nuclear contents for the therapeutic work are already present in the very beginning and in the first utterances of the client (Andolfi, 1979; Salzberger-Wittenberg, 1970; Stiles et al., 2006). In ultra-brief therapeutic interventions, these may form the main subject of the therapeutic work (Stiles et al., 2006). Likewise, family psychotherapists of all schools emphasise the role of the same initial stage of therapy (Haley, 1976; Laitila, 2004). Haley (1976) outlined a detailed structure for the first session with clearly defined stages. According to Haley, at the beginning of the very first session, the therapeutic situation and context and the roles of the participants are to be defined, and the family is introduced to the ways of working in family therapy. Rober (personal communication) has suggested writing down the first verbal utterances of each therapy session in order to come back to these later during the therapy. The common feature for many of these authors has also been recognising the fact that the meanings of the first utterances are uncovered only gradually during the course of therapy: the therapist or therapy team has no means of understanding or grasping the multitude of meanings of these opening phrases. I regard family therapy, as well as all the psychotherapies, as a context for

A. Laitila (✉)
University of Eastern Finland, School of Educational Sciences and Psychology,
Joensuu, Finland
e-mail: aarno.laitila@uef.fi

© Springer International Publishing Switzerland 2016
M. Borcsa, P. Rober (eds.), *Research Perspectives in Couple Therapy*,
DOI 10.1007/978-3-319-23306-2_4

new openings in the lives of the therapy clients, and thus the shifts towards these new openings are worth studying.

Even though the significance of this phenomenon, i.e. the importance of the beginning of therapeutic process has apparently become a part of psychotherapy lore (Laitila, 2004; Stiles et al., 2006), studies focussing on the beginning stages of psychotherapy have been rare. Shotter (2007, p. 28) referred to Goolishian's phrase for brief therapeutic work: '*if you want therapy to proceed quickly, then you must go "slow"*.' This can be seen to refer to therapists' awareness of the possible need and pressure for quick progress, a quick definition of the therapeutic goals and possible quick diagnostic understanding. All these can have a role in constructing an anti-therapeutic atmosphere with the urge to provide help as soon as possible, while the interactive possibilities are thus ignored. Already Haley (1976) recognised the need for this, and his solution was a clearly defined role of the therapist as the 'conductor' of the session and the process.

Since the days of the formation of the different schools of family therapy, evolution has been rapid. There have emerged both conversational approaches—such as reflective and narrative therapies, both of which emphasise collaboration and co-construction, semantic and semiotic work, and dialogue in a client-centred process (Anderson, 2001; Friedman, 1995; Smith & Nylund, 1997)—and more structured and manualised brief interventions, such as emotion focussed therapy (EFT) with family and couple therapeutic approaches (Johnson & Talitman, 1997; Makinen & Johnson, 2006; Sprenkle, Davis, & Lebow, 2009). Regardless of the orientation or the approach, therapists still have to decide and evaluate for themselves how interventive, active, or restrained they will be or aim to be (Minuchin, Lee, & Simon, 1996; Rober, Elliott, Buysse, Loots, & De Corte, 2008; Seikkula & Trimble, 2005).

This emphasis on the beginning stages of therapy has relevance for psychotherapy research, since ultra-brief interventions have not yet been a focus of much research. The average lenght of a process of family therapy has shortened throughout the existense of family therapy and the evolution of its separate schools, even though these are not structured ultra-brief interventions.

The purpose of the present case study is to look at the collaborative formation of interactive patterns and what options for therapeutic work emerge through them. The overall approach of the case study is a microanalytic one that tries to take a closer look at the beginning phases of the therapy process and of each session. This is done in order to shed light on ways in which therapists may respond in a sensitive situation and on the options for creating continuity of the process in order to support the consolidation of therapeutic gains. It is not only the beginning of the therapy process that is of interest, but also the beginning of each of the four constitutive sessions. The questions posed are as follows: (1) What interaction patterns are formed at the very beginning of this course of couple therapy, and how are these patterns present and evolve in the beginning phases of each consecutive session? Do these first interaction patterns have common features with the progress evaluation tool used in therapy? (2) How does the therapist use these interactions in order to introduce novelty and promote new ideas and perspectives in the existing problem story?

I shall give an account of the central interaction frames co-constructed at the beginning of the first sessions. The therapists' actions within these frames are then studied from the perspective of the four-session therapy process; thus, after the presentation of the interaction frame, it is also evaluated in the four-session context. The aim is to achieve a comprehensive picture of the in-session interaction patterns and to compare these across the entire process with emphasis on the therapists' perspective.

Theoretical Review

Therapists' actions in family therapy have been bound and rooted in different schools of family therapy (Dallos & Draper, 2000; Goldenberg & Goldenberg, 2013; Gurman & Kniskern, 1991; Lebow, 2005). This way of thinking about the therapist's role emphasises the differences that arise from different theoretical orientations. Sprenkle et al. (2009) term these differences as *model driven*. Other possible approaches for examining therapists' actions are, for example, interest in therapists' ways of asking questions in different modes of therapy (McGee, Del Vento, & Beavin Bavelas, 2005), the contextual model (Wampold, 2001), the common factors approach (Sprenkle et al., 2009), or the expertise-driven method (Laitila, 2004; Seikkula, Arnkil, & Erikson, 2003). Rather than relying on model-driven considerations, these ways of looking at therapeutic action rather focus on the commonalities of multi-actor therapeutic settings and ways of promoting therapeutic conversation rather than just model-specific features.

The contextual meta-model, common factors, and expertise-driven approaches highlight the basic qualities that specify and differentiate psychotherapy in general, and family therapy in particular, from other social encounters. However, these basic qualities are seen as common to different therapy modes. On the other hand, the model-specific perspective on psychotherapy (or the medical meta-model) emphasises variable-centred ways of differentiating the key factors of therapeutic change (Wampold, 2001). These variables are assumed to include the effective ingredients that are specific to each therapy mode.

Nowadays, the role of the beginning of the psychotherapy process is not just a question of professional lore, as research has justified an emphasis on the beginnings of the sessions. Already, the first utterances of the client and way in which these are jointly dealt with have been shown to be significant (Stiles et al., 2006). Important factors include the work towards achieving collaboration and the formation of a therapeutic system (Laitila, Aaltonen, Wahlström, & Angus, 2001), work towards establishing and consolidating therapeutic gains and securing continuity of the process (Gorell Barnes, 2004), focus on collaboration, therapeutic alliance, client agency (Kuhlman, 2013), and the significance of client feedback for the therapeutic process right from its outset (Lambert, 2013; Lambert, Whipple, Vermeersch, Nielsen, & Hawkins, 2001).

Evaluation methods used during the course of therapy—such as Clinical Outcome in Routine Evaluation (CORE-OM), Outcome Rating Scale (ORS), Session Rating Scale (SRS), and Outcome Questionnaire-45 (OQ-45)—have provided both clinicians and researchers with accessible tools and ways of assessing progress in therapy. These measures consist of a sample of questions (CORE-OM, OQ-45) or visual–analogic scales (ORS, SRS) with which to evaluate the course of a therapy session or its outcome. A common feature of these is that the measures connected with the therapeutic change are given prior to a session or in the beginning of a session. These measures have highlighted the importance of client experience and feedback and the significance of these for the clinical outcome (Lambert, 2013; Lambert et al., 2001; Sparks, Kisler, Adams, & Blumen, 2011). Sparks et al. (2011) showed how SRS and ORS, while providing therapists with significant client feedback, can be applied to promote therapists' professional development. In addition, Kuhlman (2013) carried out research on clients' activity in assessing the progress of marital therapy for depression.

Early response in treatment has been interpreted according to the pharmacological research as an evidence of the placebo effect (Quitkin, McGrath, Stewart, Taylor, & Klein, 1996; Stewart et al., 1998), but Lambert (2005) has criticised this interpretation, both on empirical grounds and on grounds of theoretical problems with the concept of a placebo in the context of psychotherapy research. Early response has been interpreted to speak in favour of common factors of psychotherapies on the one hand (Lambert, 2005) and for client qualities on the other. Undoubtedly, the early therapeutic response will be a question of interest in the future as well, as more structured and manualised brief interventions are developed.

There are qualitative differences between dialogues in individual and family settings. This is the case even though Schielke et al. (2011) emphasise the isomorphic relationship between the intrapersonal and interpersonal processes of couple therapy. Research informed by system theory has revealed two different modes of communicating in human interaction: analogical and digital communication, first of which is related to relationship aspect of communication, and understood in context, and the latter of these is related to the content aspect with meaning that is unambiguous and agreed upon. Thus, there is a continuous reciprocal process defining the relationships between the participants (Watzlawick, Beavin, & Jackson, 1967). In family therapy, as in individual psychotherapy, there are both inner and outer conversations taking place. In family therapeutic conversations, the most significant persons that are talked about (and who often participate in the definition of the problem) are present, not as internal objects, or inner alters (Markova,2006), but as living persons. We can describe this as the multiplicity and multidimensional character of ongoing family therapeutic conversations or the dialogical nature of family therapeutic interaction. The participants actively position themselves at different points in the interaction using different voices according to their present position in the conversation and the anticipated answer (Rober et al., 2008; Seikkula, Laitila, & Rober, 2012). Dialogue in the 'here and now' is emphasised in most family therapeutic orientations. All participants try to accommodate their own expressions, both as to what is said and as to what is expected. While speaking to or answering the therapist, participants are aware of the fact that the therapist is not the only one present to hear their speech. Conceptualisations of difficulties in the context of family and marital therapy take place in relationship terms (Sprenkle et al., 2009). These qualitative differences call for different tools of research than those of individual psychotherapy research. It is possible to regard the dyadic setting of individual psychotherapy as a community of interactively communicating voices (Stiles, Honos-Webb, & Lani, 1999), but the social interaction that occurs in a multi-actor setting also calls for tools to look at and listen to these multiple participants as living persons in the situation, not just as voices. In this chapter, as in the other chapters of this volume, I shall present one research approach that is specifically tailored for use in multi-actor settings and illustrate its application with a case study.

Materials and Methods

This case study takes as its object of analysis the beginning phases of each therapy session—the first speech turns of each session, which may consist of greetings, filling out the ORS as the progress measure, and the first meaningful therapeutic topics. The sections of dialogue to be analysed were made according to the synopsis by Wahlström (see Chap. 10), and the aim of the selection was to emphasise the beginning of the therapy session and the emerging therapeutic conversation in each session. The definition of 'the beginning' included both the formalities of entering the therapy room and also the first initiatives, by any member of the therapeutic system, towards therapeutic dialogue and responses to these. The assumption was that, in the course of filling out the ORS and making the first speech turns, some continuity of process is constructed, with optional references to previous conversations, to the overall therapeutic process and to possible changes that have taken place. Selections of dialogue were given titles as follows, in chronological order: (1) Initial problem formulation (session 1, turns 7–41) and difference between 'then' and 'now' (session 1, turns 42–70); (2) exploring the present situation (session 2, turns 1–77); (3) exploring the present situation (session 3, turns 1–80); and (4) exploring the current situation and reasons for change (session 4, turns 1–102). These titles are abstractions representing the data and do not contain client utterances or direct quotations of transcripts.

The analysis was done through a microanalytical reading of session transcripts using Seikkula et al.'s (2012) dialogical toolkit, Dialogical Investigations in the Happenings of Change (DIHC), in the context of the application of the narrative process modes and concepts of the Narrative Processes Coding System (NPCS) developed by Angus, Levitt, and Hardtke (1999). The NPCS methodology has been applied in a more microanalytic style in multi-actor settings, as was the original model of Angus and her associates (Laitila et al., 2001; Laitila, Aaltonen, Wahlström, & Angus 2005; Vall, Seikkula, Laitila, & Holma, 2014, 2013).

The term 'narrative process modes' comes from the research tradition of individual psychotherapy process research (Angus et al., 1999; Laitila et al., 2001). The term is used in this chapter to refer to 'ways or modes of inquiry or cognitive/affective processes through which (a) clients strive to understand themselves and (b) clients and the therapists co-construct understanding during therapy sessions' (Laitila et al., 2001, p. 310) (these ways or modes of inquiry are described in detail during the description of the STEP III of analysis of this case study). While developing the NPCS methodology, according to Angus (personal communication), the idea was to find a pragmatic tool that would allow both quantitative and qualitative research approaches.

The research method applied in this case study is one of the several alternative ways to conduct DIHC (Seikkula et al., 2012; Chap. 5) and includes three separate steps. After exploring topical episodes (STEP I) and exploring the series of responses to the utterances (STEP II), the third step in exploring the narration (or the language area) is carried out with the tools provided by the NPCS (Angus et al., 1999; Laitila et al., 2001; Seikkula et al., 2012) (STEP III).

For STEP I, the division into topical segments in the synopsis met the needs both for the DIHC (Seikkula et al., 2012) and the NPCS (Angus et al., 1999; Laitila et al., 2001, 2005). The topic selection was made with both DIHC and the NPCS so that one topical episode or topic segment forms a coherent passage dealing with a singular theme. The second and third steps of the analysis provide more specified and focussed ways of looking at chosen episodes.

During STEP II, the analysis was applied so that the therapeutic conversation was first read with attention to the *addressees* to whom the speaker is directing her or his utterances, and the *voices* he or she is using (see Chap. 5) other than her or his own in these utterances. The question of addressees was usually easier or simpler one to solve just by looking at the transcripts. The therapist might say, for example, 'How can I help you?' In a multi-actor setting, with this utterance the therapist is offering the speech turn to the clients without specifying who should be the one to respond. One of the clients might answer: 'We discussed this and could not really agree, but I must say…'. Here the client is referring both to the 'we' as a collective voice and to 'I' as a personal voice. As Seikkula and his associates wrote, 'a single utterance by a single participant can include many voices' (2012, 10).

The emphasis of STEP III was on small shifts in narration by the participants. Attention was first given to the tone of speech, utterances, co-construction of meanings, and possible shifts in these. Then we proceeded with a microanalytic reading of narrative process types (Angus et al., 1999; Laitila et al., 2001; 2005; Seikkula et al., 2012). The aim was to study how the speaker processes the contents of her or his expressions in the context of a specific topic. This can be done with (1) the external process mode as descriptions of events (either real or imagined ones); (2) the internal process mode as experiential descriptions of feelings and inner states, for example, by using emotion descriptions or metaphoric expressions; or (3) the reflexive mode as efforts to understand the meaning of both happenings and inner states (Angus et al., 1999; Laitila et al., 2001, 2005). For example, (1) the description of an event by the client could be an account like 'Today we had to take a taxi to reach this session in time'; (2) an experiential description might follow, saying that 'and being in this kind of hurry makes me real tensed, just like a balloon with a little too

much gas in it'; then (3) a reflexive evaluation may be given of the possible consequences and meaning of the described events, for example, 'I have sometimes won-
dered that why is it so. Have I learned this way or attitude from my grandfather who
used...?' Through this microanalytic reading, I have examined the different offer-
ings and responses of the participants in their joint efforts to make meanings in the
therapy dialogue.

Findings

The research task was to look at interaction patterns as these are constructed at the
beginning of the therapy process and how these patterns are used in the course of
therapy. Four interaction patterns were identified. In this section, I give an account
of these four patterns, followed by examples of each. The four patterns are identified
as follows:

1. Responsive, nonchallenging joining, and attunement by the therapist to the client
   couple's interactive approach characterised by therapist's usage of both clients'
   words, open questions, repeating, or minor rephrasing of clients' utterances
2. Metaphoric shifts by the participants
3. Joint actions of the partners to present the couple's relationship (as the scene of
   joint understanding) while introducing oneself to the therapist
4. The therapist's moves to organise the therapeutic conversation, this consisting of
   different active ways of asking questions or ways of sharing speech turns

At the beginning of the first session, all four interaction patterns were constructed
and present, and traces of three of these could be found in each of the four sessions.
All these styles called for new ways of participating in the discussion, offering both
alternative and restricted options.

*1. Responsive, nonchallenging joining, and attunement by the therapist with the
client couple's interactive approach*

Extract 1

Session 1, turns 7–9

7 T		OK (.) so where would you start?
8 V		We can start from the beginning [giggles], it's , the situation is that, it was summer, [2 years earlier] that I started, getting , I got depressed , and now now my situation is better (.) it's been like half a year that I've been fine. I'm still eating medication but already like in smaller portions.
9 T		Yeah

The therapist's question 'where would you start' (1:7) offered an open-ended
option for the clients to give an account of their reasons for seeking help for themselves. In Victoria's response, referring to the beginning, the first part of her answer
was open-ended in a similar fashion to the therapist's question. By starting her utterance in this way, she gave an ambiguous answer and referred at least to the beginning

of the relationship, the depression, and the difficulties in the relationship with an open-ended timeframe. The ambiguous richness of the expression seemed to be amusing or a bit baffling (or both) for herself, too, maybe because the literal answer also had an ironic or even hostile overtone in it.

Initially, the therapist was attuned to Victoria's overall speech more than single utterances or words, thus encouraging Victoria to use her own ways of describing their reasons for coming to therapy. For Victoria, it seemed to be essential to elaborate 'the beginning' and its meaning for her, being the one to provide an account or external description. Thus, she could take the central responsibility for seeking therapeutic help for both of them.

This nonchallenging responsivity of the therapist was present during nearly the whole initial problem formulation (turns 7–41); as a kind of minimalistic therapeutic intervention, it continues throughout the four sessions.

2. The metaphoric shifts by the participants

Extract 2

Session 1, turn 10
10 V   but the reason why we are here is that this, my thing left kind of scars in our relationship because when I was down, when I was feeling bad, it was really (.) baad, it affected, affected a lot in our relationship (text omitted)

After she referred to her depression as 'the beginning', Victoria said 'but' (1:10), which highlighted her differentiation of the treatment process of her own depression from the present difficulties of her relationship with Alfonso. She described the earlier treatment process and issues connected with it as 'my thing left kind of scars', and externalised her depression as something that had certain effects for their mutual relationship, not only for herself and her subjective experiences.

The 'scars in our relationship' was the first metaphoric expression to be used at the very beginning of the first session. This metaphor was both a precise and an open-ended definition, referring to psychological pain in the situation the way that Victoria saw it affecting both of them. 'Scar' as a metaphor also referred to something that can either call still for treatment or as a sign of previous struggles, but has already healed enough. By using the metaphor, Victoria moved the conversation into the relationship focus, inviting Alfonso to participate in responding to the therapist's invitation to define the situation somehow.

At the beginning of the second session, the clients fill out the ORS form for the first time and there is discussion of the meanings of the questions.

Extract 3

Session 2, turn 16
16 T1   and what do the signs tell about (.) your life at the moment?

While referring to the ORS methodology, the therapist also asks metaphorical questions, for example, '...and what do the signs tell about (.) your life at the moment?' His question calls for personal interpretive accounts and self-disclosure, since the marks on the lines of the ORS do not 'tell' anything without the speakers' own words. This kind of metaphoric asking is present in both sessions 2 and 3, when the outcome measure is dealt with.

Both of these first two patterns—(1) the therapist's responsive nonchallenging style and (2) the use of metaphoric communication by all the participants of the interaction—served to open up more conversation by accepting the clients' own expressions and thus inviting and encouraging them to give more descriptions and new perspectives. Both examples of metaphoric communication—'scars' in excerpt 2 and 'signs telling' in excerpt 3—are open-ended and allow for use of language that is both concrete and behavioral in thus externalising or verbalising intimate personal experiences.

3. Joint actions of the partners to present the couple's relationship while introducing themselves to the therapist

Extract 4

Session 1, turns 14–18

14 A		Yeah, but you, yes, the thing it's just that [clears his throat] after this time now, that, that, if we can say that she's a lot better it means that she feels better now but still when we have this kind of situations that, earlier I felt I was aa I had kind of more patience, I could, I could go, like how'd you say (.) I had more patience in listening
15 V		In comforting me
16 A		Yes, yes, now I feel that it's like I don't, somehow even if there is some small thing
17 V		Some small thing that I would like to discuss about, for example
18 A		But then I think that I, somehow I react too strongly like if it would be something bigger

After the external-type account of 'the beginning', in which Victoria connected life situation, life events, and the challenges with the depression and internal-type elaboration of some dimensions of personal experiences with Victoria's talk of 'the scars', both partners entered into a more evaluative and reflexive conversation of the meaning of the present situation for both of them and for their relationship.

This joint communication, completing each other's expressions and continuing each other's sentences, formed their 'dance' inside and around the meanings of the situation. It highlighted their (1) consensual agreement about the situation; (2) difficulties of differentiation in the couple relationship, and the possible threat to the relationship due to differentiation; (3) developmental phase of their intimate relationship, in connection to (2); and (4) explicit account to convince the therapist why they needed psychotherapy. All of these factors represented optional openings into their landscape of meaning (Carr, 1998; White, 1995), and all were accomplished during the first 18 turns of the first session.

In the second session, this frame was further developed while the partners began evaluating their ORS scores, and the differences between their personal situations became evident.

Extract 5

Session 2, turns 29–32
29 V	I didn't have time to sleep or (.) and I just feel a bit sad, because of this thing, I don't know why we feel so differently
30 T1	mm (.) you are sad because of these differences on [points to the paper]?
31 V	No, I am happy that he feels good
32 T1	OK, OK, yeah, yeah (.) but you are sad, you feel sad

Victoria wanted to make a distinction between her sadness (or not knowing the reason for it) surrounding their different views on the situation and that surrounding her own internal state of mind. This was further clarified by the therapist's declarative statement. In being attuned to the internal process mode of Victoria in his own response, the therapist had a role in her dealing with her personal experience.

4. The therapist's moves in organising the therapeutic conversation

Extract 6 (clients' responses omitted)

Session 1, turns 19–37
19 T1	So you are saying that meanwhile Victoria was down you could support her but now when things are better, in some way,}
	(…)
23 T1	Mmm. So that was the case when you was down and you were always sleeping and crying in the bed for 2 days.
	(…)
30 T1	You are afraid that it can be again a similar situation as it was
	(…)
37 T1	Do you have some example in your mind that's happened in the last days for example

This extract shows how the therapist's moves took place as small interventions, such as:

(a) Sharing the comments and questions for both the partners addressing what is said
(b) Validating their accounts and experiences
(c) Constructing brief summaries with an interrogative tone, or repeating and rephrasing what was said earlier
(d) Active asking, which was a therapist's move that was not present at the beginning of the first session. This had somewhat separate tones in the second and third sessions and in the fourth session

As it turned out that Victoria was more talkative than Alfonso, the therapist, in addition to just asking open questions, explicitly addressed Alfonso with questions. This inevitably emphasised the therapist's role in organising the ongoing conversation. The most interventive move of the therapist was to ask for an example (1:37); this move shifted the therapist even more into the central position.

Active asking, in the second and third sessions, consisted of open questions and comments made in an interrogative manner, for example, repeating literally, paraphrasing or adding some of the therapist's own words to something that had just been said.

In the final session, however, active asking took a different form. It emphasised differences, change and outcome, empowerment and resource orientation as compared to the first session and sessions subsequent. The focus on resources and the positive tone were present serving the consolidation and stabilisation of the changes made, as well as crediting the couple for these, and for the continuous therapeutic process of the client couple while terminating the therapy sessions.

Through a closer look at these first turns, concerning the dominant interaction styles, we were able to see how much content information and relationship information were present. The participants were attuning themselves to the anticipated answers of the other persons present. Thus, they were showing interactive competence in their positioning in these beginning phases of the therapeutic sessions and therapy process, and thereby leaving the stage open for the possibility to choose the level of intimacy and trust both for themselves and each other.

There were two clear instances of variation from the above-mentioned patterns of moves by the therapist. These have been considered separately from the other moves due to their rarity and also their challenging nature. The first one occurred in the first session, as the therapist asked for an example from Alfonso (1:37). This came just after Alfonso had tried to give an account of his own perspective (1:36). The therapist invited him to be more concrete, in effect, challenging him. As this was done at the beginning of the first session, when the working relationship was still delicate, it could be regarded as taking a risk on the part of the therapist. However, in addition to being risky, it also had a joining or supportive nuance, since the therapist wanted to hear Alfonso's perspective after Victoria's long speech turn. For Alfonso, from the process perspective, it seemed that he experienced the question as a resource. The therapist's suggestion in turn 37 seemed to provide him with a way to express himself and in the later sessions he used examples of himself in order to tell his story and explain his view.

The second variation was the way in which the therapist challenged Victoria in the second session to give an account of her expression that '... if some evening I am home it's never at the same time with him' (2:58). Victoria's word 'never' in turn 58 could be understood, from a cognitive perspective, as a depression-specific overgeneralising statement (Beck, Rush, Shaw, & Emery, 1979; Beck, Steer, & Garbin, 1988). In turn 64, the therapist responded to it by asking for a more detailed and concrete description of the contents of the utterance in the relationship context. The therapist said, 'You said never, what does it mean?' (2:61) and then in turn 64 clarified his question, '. That you never meet both of you, that you are never together home, so, so when was the last time you were together home with' (2:64). This instance comes close to active asking, but the therapist's clarification in turn 64 shows that it is more than that. The therapist takes an active orientation in the direction of talking about exception of the rule of 'never', instead of just accepting the given answer.

Both of these two variations in pattern connote that the therapist kept both partners accountable, expected accountability, and showed trust in both partners' ability to speak for themselves, regardless of the emotional state of each or how tension-filled the topic at hand. This took place in a situation where Victoria had her original experience of psychotherapy for her depression and Alfonso came as a beginner in psychotherapy to the couple therapy process. By actively asking from both partners, the therapist both challenged and empowered them.

Discussion

This case study aimed to find out how novel issues are introduced to the conversation in the couple therapy process. This was done by taking a closer look at the interaction patterns taking place at the beginning of each session. Four distinct patterns were recognised. According to the findings, no single pattern of interaction was dominant, or played a key role in introducing novelties or promoting change as such.

Since the nonchallenging 'therapeutic style' associated with joining in with clients' utterances is quite a restrained one, it makes the instances of exception to this style interesting. These exceptions easily punctuate the dialogue and highlight what should be noted by the clients. Besides this responsive nonchallenging joining, it became obvious that the minor shifts by the therapist—such as more active asking than simply rephrasing or repeating what was said, either in an interrogative tone or in a more neutral tone—became remarkable, since these positioned him in a more central role in the therapeutic interaction. While therapists are able to move 'to and fro' in this sense—closer and more distant—they cannot remain neutral. Rather, they must actively position themselves as the agents for change. This regulation of distance carried out with minor shifts calls for a more detailed analysis in different multi-actor settings. The marital therapy setting emphasises the collaboration of adult actors, and it would be interesting to look at the same issue in settings with different family compositions, especially in families with children of different ages.

In the context of the generally nonchallenging atmosphere of conversation, the power of asking questions and the different ways of doing so became evident in our analysis. As our analysis showed, asking for an example can be either a supportive, dialogue-encouraging attitude, or a criticism aimed towards a given account. During the course of therapy, the former seemed a more plausible option; in later sessions, the client could use the very same expression the therapist had originally used. In the final session, it seemed that the male therapist had adopted a completely new agenda for his way of asking questions, as they became resource-oriented, focussing on the outcome and the changes that had taken place. This is similar to the phenomenon Partanen (2008) found in her detailed analysis of therapists' talk in follow-up sessions of therapy groups for men who had behaved violently in their intimate relationships. The therapists' agenda is one of the consolidating changes made during the treatment process and follow-up period, and co-constructing the success story arising from the details that the clients brought up.

The theoretical differences between individual psychotherapy and multi-actor therapeutic interaction were emphasised, especially the need for analytical and theoretical tools developed specifically for multi-actor dialogues. Relationship difficulties are dealt with in the presence of living others instead of inner alters. In this case, the therapist had to invite the more passive partner to participate. It is also noteworthy that Victoria constructed an invitation for Alfonso to participate, so that therapeutic conversation would involve both of the partners. Couple and family therapeutic conversation is a context for the common factors to work for the therapeutic aims, but the common factors of individual psychotherapy and its dyadic setting are not enough to illuminate the process of fitting and positioning oneself in a multi-actor interaction. The object of a speech is generally shared by all the members of the client group and is not necessarily regarded purely as an intrapsychic reality or construct. The participants in the interaction act both as speakers in their own right, addressees, witnesses, and participants in the therapy talk. This cannot be covered by the analytic tools designed for dyadic therapy contexts, which have different basic assumptions about the production of psychological states. The 'living alter' is different from the 'inner alter' in therapeutic practice, and research methods cannot ignore this fact (Markova,2006; Rober, 2005; Sprenkle et al., 2009).

Methodologically, the application of narrative process modes as a part of dialogical analysis provides openings both in the meaning-making and experiential spheres of the clients, as well as in the nuances of therapists' points of joining in the interaction and in the conversation. While working with nonchallenging interventions, therapists can still simply call on the clients to provide further accounts and simultaneously encourage them to move forward in the descriptions of their experiences and their understandings of these.

Two issues still merit further research. As well as setting up positive interactional parameters, the beginning stages may also play a role in possible antitherapeutic developments; it would be useful to find the potentially detrimental and noncurative factors, the effect of which could not be corrected in the course of the therapeutic process, that may be established in this period. This analysis also could not show the effect of the variable of clinical experience and possible differences connected with this in the utterances and reflections of the therapists, the second therapist being a student. The difference in the amount of experience between the two therapists was large, but the role of the second therapist was too minor in the data of this case study to justify any valid conclusions.

Just as the therapeutic approach, method or technique in itself cannot perform the therapy without actors in a position of agency; similarly, the research method does not carry out the research process. Both call for living persons to carry out this process. Hence, it is evident that, in using qualitative methods, the researcher has also used empathy as a research strategy and as a key factor in carrying out the research. Accordingly, the observations made are contextual in character (Elliott, 2012; Stiles, 1993, 2003), and there remains a degree of interpretation in these results. The reported findings are grounded in the data and their extraction is made transparent using the examples of transcribed conversations. Still, this does not mean that the thrust of the interpretations could not be challenged by an analysis by some other researcher.

Conclusion

Certain interaction patterns were already co-constructed in the first session, and the formation of the first four patterns was already visible during the first two topics of the first session. Other patterns may have emerged in the rest of the sessions, but the analysis in this case study has confined itself to the beginning phases of each session. The methodology allowed for attention to both the contents of the conversation and the ways different participants made themselves understood and heard. Concerning the therapist's role, the conclusion for family therapy practice favours the therapeutic presence of careful empathetic listening and a courageous attitude towards exploring the details of the participants' speech.

References

Anderson, H. (2001). Postmodern collaborative and person-centered therapies: What would Carl Rogers say? *Journal of Family Therapy, 23*, 339–360.

Andolfi, M. (1979). *Family therapy: An interactional approach*. New York: Plenum Press.

Angus, L., Levitt, H., & Hardtke, K. (1999). The narrative processes coding system: Research applications and implications for psychotherapy practice. *Journal of Clinical Psychology, 55*, 1250–1270.

Beck, A. T., Rush, S. J., Shaw, B. F., & Emery, G. (1979). *Cognitive therapy of depression*. New York: Guilford Press.

Beck, A. T., Steer, R. A., & Garbin, M. G. (1988). Psychometric properties of the Beck Depression Inventory: Twenty-five years of evaluation. *Clinical Psychology Review, 8*, 77–100.

Carr, A. (1998). Michael White's narrative therapy. *Contemporary Family Therapy, 20*, 485–503.

Dallos, R., & Draper, R. (2000). *An introduction to family therapy. Systemic theory and practice*. Buckingham: Open Universities Press.

Elliott, R. (2012). Qualitative methods for studying psychotherapy change processes. In D. Harper & A. R. Thompson (Eds.), *Qualitative research methods in mental health and psychotherapy* (pp. 69–81). Oxford: Wiley-Blackwell.

Friedman, S. (Ed.). (1995). *The reflecting team in action. Collaborative practice in family therapy*. New York: Guilford Press.

Goldenberg, H., & Goldenberg, I. (2013). *Family therapy. An overview* (8th ed.). Belmont, CA: BrooksCole.

Gorell Barnes, G. (2004). *Family therapy in changing times* (2nd ed.). Bristol: Palgrave McMillan.

Gurman, A. S., & Kniskern, D. P. (Eds.). (1991). *Handbook of family therapy* (Vol. II). New York: Routledge.

Haley, J. (1976). *Problem-solving therapy*. San Francisco, CA: Jossey-Bass.

Johnson, S. M., & Talitman, E. (1997). Predictors of success in Emotionally Focused Marital Therapy. *Journal of Marital and Family Therapy, 23*, 135–152.

Kuhlman, I. (2013). *Developing accountability in couple therapy for depression: A mixed methods study in a naturalistic setting in Finland* (Jyväskylä studies in education, psychology and social research, Vol. 482). Jyväskylä: Jyväskylä University Printing House.

Laitila, A. (2004). *Dimensions of expertise in family therapeutic process* (Jyväskylä studies in education, psychology and social research, Vol. 247). Jyväskylä: Jyväskylä University Printing House.

Laitila, A., Aaltonen, J., Wahlström, J., & Angus, L. (2001). Narrative process coding system in marital and family therapy: An intensive case analysis of the formation of a therapeutic system. *Contemporary Family Therapy, 23*, 309–322.

Laitila, A., Aaltonen, J., Wahlström, J., & Angus, L. (2005). Narrative process modes as a bridging concept for the theory, research and clinical practice of systemic therapy. *Journal of Family Therapy, 27*, 202–216.

Lambert, M. J. (2005). Early response in psychotherapy: Further evidence for the importance of common factors rather than 'placebo effects'. *Journal of Clinical Psychology, 61*(7), 855–869.

Lambert, M. J. (2013). Outcome in psychotherapy: The past and important advances. *Psychotherapy, 50*(1), 42–51.

Lambert, M. J., Whipple, J. L., Vermeersch, D. A., Nielsen, S. L., & Hawkins, E. J. (2001). The effects of providing therapists with feedback on patient progress during psychotherapy: Are outcomes enhanced? *Psychotherapy Research, 11*(1), 49–68.

Lebow, J. L. (Ed.). (2005). *Handbook of clinical family therapy*. Hoboken, NJ: Wiley.

Makinen, J. A., & Johnson, S. M. (2006). Resolving attachment injuries in couples using emotionally focused therapy: Steps toward forgiveness and reconciliation. *Journal of Consulting and Clinical Psychology, 74*(6), 1055–1064.

Markova, I. (2006). On 'the inner alter' in dialogue. *International Journal for Dialogical Science, 1*, 125–147.

McGee, D., Del Vento, A., & Beavin Bavelas, J. (2005). An interactional model of questions as therapeutic interventions. *Journal of Marital and Family Therapy, 31*, 371–384.

Minuchin, S., Lee, W.-Y., & Simon, G. S. (1996). *Mastering family therapy. Journeys of growth and transformation*. New York: Wiley.

Partanen, T. (2008). *Interaction and therapeutic interventions in treatment groups for intimately violent men*. Jyväskylä: Jyväskylä University Printing House.

Quitkin, F. M., McGrath, P. J., Stewart, J. W., Taylor, B. P., & Klein, D. F. (1996). Can the effects of antidepressants be observed in the first two weeks of treatment? *Neuropsychopharmacology, 15*, 390–394.

Rober, P. (2005). Family therapy as a dialogue of living persons: A perspective inspired by Bakhtin, Voloshinov, and Shotter. *Journal of Marital and Family Therapy, 31*, 385–397.

Rober, P., Elliott, R., Buysse, A., Loots, G., & De Corte, K. (2008). Positioning in the therapist's inner conversation: A dialogical model based on a grounded theory analysis of therapist reflections. *Journal of Marital and Family Therapy, 34*, 406–421.

Salzberger-Wittenberg, I. (1970). *Psycho-analytic insight and relationship: A Kleinian approach*. London: Routledge.

Schielke, H. J., Stiles, W. B., Cuellar, R. E., Fishman, J. L., Hoener, C., Del Castillo, D., et al. (2011). A case study investigating whether the process of resolving interpersonal problems in couple therapy is isomorphic to the process of resolving problems in individual therapy. *Pragmatic Case Studies in Psychotherapy, 7*, 477–528.

Seikkula, J., Arnkil, T. E., & Erikson, E. (2003). Postmodern society and social networks: Open and anticipation dialogues in network meetings. *Family Process, 42*, 185–203.

Seikkula, J., Laitila, A., & Rober, P. (2012). Making sense of multi-actor dialogues in family therapy and network meetings. *Journal of Marital and Family Therapy, 38*, 667–687.

Seikkula, J., & Trimble, D. (2005). Healing elements of therapeutic conversation: Dialogue as an embodiment of love. *Family Process, 44*, 461–475.

Shotter, J. (2007). Not to forget Tom Andersen's way of being Tom Andersen: The importance of what 'just happens' to us. *Human Systems: The Journal of Systemic Consultation & Management, 18*, 15–34.

Smith, C., & Nylund, D. (Eds.). (1997). *Narrative therapies with children and adolescents*. New York: Guilford Press.

Sparks, J. A., Kisler, T. S., Adams, J. F., & Blumen, D. G. (2011). Teaching accountability: Using client feedback to train effective family therapists. *Journal of Marital and Family Therapy, 37*, 452–467.

Sprenkle, D. H., Davis, S. D., & Lebow, J. L. (2009). *Common factors in couple and family therapy*. New York: Guilford Press.

Stewart, J. W., Quitkin, F. M., McGrath, P. J., Amsterdam, J., Fava, M., Fawcett, J., et al. (1998). Use of pattern analysis to predict differential relapse of remitted patients with major depression during 1 year of treatment with fluoxetine or placebo. *Archives of General Psychiatry, 55*, 334–343.

Stiles, W. B. (1993). Quality control in qualitative research. *Clinical Psychology Review, 13*, 593–618.

Stiles, W. B. (2003). Qualitative research: Evaluating the process and the product. In S. Llewelyn & P. Kennedy (Eds.), *Handbook of clinical health psychology* (pp. 477–499). Chichester, UK: Wiley.

Stiles, W. B., Honos-Webb, L., & Lani, J. A. (1999). Some functions of narrative in the assimilation of problematic experiences. *Journal of Clinical Psychology, 55*, 1213–1226.

Stiles, W. B., Leiman, M., Shapiro, D. A., Hardy, G. E., Barkham, M., Detert, N. B., et al. (2006). What does the first exchange tell? Dialogical sequence analysis and assimilation in very brief therapy. *Psychotherapy Research, 16*, 408–421.

Vall, B., Seikkula, J., Laitila, A., Holma, J. & Botella, L. (2014). Increasing responsibility, safety, and trust through a dialogical approach: A case study in couple therapy for psychological abusive behavior. *Journal of Family Psychotherapy, 25*, 275–299.

Vall, B., Seikkula, J., Laitila, A., & Holma, J. (2013). Complexities of dialogue in therapy for intimate partner violence. Addressing the issue of psychological violence. Submitted.

Wampold, B. E. (2001). *The great psychotherapy debate. Models, methods, and findings*. Mahwah, NJ: Lawrence Erlbaum.

Watzlawick, P., Beavin, J. H., & Jackson, D. D. (1967). *Pragmatics of human communication: A study of interactional patterns, pathologies, and paradoxes*. New York: Norton.

White, M. (1995). *Re-authoring lives*. Adelaide: Dulwich Centre.

Chapter 5
Therapists' Responses for Enhancing Change Through Dialogue: Dialogical Investigations of Change

Jaakko Seikkula and Mary Olson

The point of view on research represented here is based on a dialogical framework. Our emphasis is on understanding the contribution of the therapists to the process of change. We will examine the first three couple therapy sessions with Alfonso and Victoria.

The Notion of Dialogue in Psychotherapy Practice and Research

Psychotherapy can be understood as a process of listening and finding words for previously unspeakable or inexpressible experiences. The responses of the therapist(s) in the exchange with the family are key ingredients in the creation of a new and common language for the person's distress that otherwise remains embodied and expressed in symptoms. In family therapy, family members as real living persons in the actual session become invaluable participants in the search for new forms of expression, in which new possibilities for meaning and action reside. Following Mikhail Bakhtin's (1984) concept of dialogicality, the responsiveness of the therapist and the presence of family members are very important, because interlocutors, those who take part in the conversation, are active co-authors of a person's utterances and meanings.

J. Seikkula (✉)
University of Jyväskylä, Jyväskylä, Finland
e-mail: jaakko.seikkula@jyu.fi

M. Olson
University of Massachusetts Medical School, Worcester, MA, USA

Smith College School for Social Work, Northampton, MA, USA

© Springer International Publishing Switzerland 2016
M. Borcsa, P. Rober (eds.), *Research Perspectives in Couple Therapy*,
DOI 10.1007/978-3-319-23306-2_5

We conceptualize therapeutic conversation as a dialogical activity (Olson et al., 2011; Rober, 2005b; Seikkula, & Arnkil, 2014). In using the research method, "The Dialogical Investigations of Happenings of Change", DIHC (Seikkula et al., 2012), the focus is on how therapists participate and answer from a specific position of "responsive responsibility" (Steiner, 1989). By this, we mean that we are part of a joint project for increasing the understanding of the specific situation in which a request for help has been made. Understanding occurs as an artifact of an active process of the therapists' answering clients' utterances, which is an act of taking responsibility for the other and for the situation. In an open dialogue, "utterances are constructed to answer previous utterances and also to wait for an answer from utterances that follow" (Seikkula, 2002, p. 268).

In this chapter our aim is twofold. First, we will present a method for conducting a dialogical analysis of couple sessions. Second, we track detailed sequences from which the events of change become discernable. To make sense of the unfolding details of this process, we use Bakhtinian concepts including, "voice," "addressee," and "positioning" which are all facets of the larger prism of dialogue. These concepts are helpful for the researcher in conducting the microanalysis of select topical episodes. First, we will define these concepts. The next section will describe how these concepts are translated into a research method. And, the third part will give an example of how this method can be used to elucidate a process of change.

Voices

Being heard and thus, having a voice occur and are witnessed in terms of how a person is responded to and whether this response invites their further participation in the dialogue. A couple or family therapy session constitutes a key context where new language, new voices, and ultimately, new stories, can develop gradually and in unpredictable ways out of the often tense interactions of everyone present (Bakhtin, 1981). The concept of "voices" is central to this process, though difficult to give a precise definition. There have been various attempts to do so. For instance, Stiles, Osatuke, Click, and MacKay (2004) offer a cognitive and ecological view of voices to operationalize the idea, noting that "Voices are traces and they are activated by new events that are similar or related to the original event" (p. 92). As Stiles et al. say, all of our experiences leave a sign on the body, but only a fragment of these signs ever become expressed as spoken narratives. Although this view is valuable, in adopting a dialogical framework, we emphasize the social and relational nature of voices. As Bakhtin (1984) notes, voices are the speaking consciousness (Wertsch, 1991). A voice becomes alive in an interchange. Voices are not "things" inside a person, but they only live in a continuous flow of interaction with others. Thus our embodied experiences formulated into words in a dialogical context become the voices of our lives.

Every utterance has an author whose position it expresses (Bakhtin, 1981, 1984). The words I utter can be my very own words, or they may include the words of other people. In this way, I speak in different voices, which allows me to take different positions toward these voices. As Bakhtin notes, small children in school already learn the polyphonic quality of language while reading aloud. Take, for instance, another example: the conversation within supervision. The therapist (supervisee) is describing what his client has said and then commenting on that. In this utterance, at least two voices are present, i.e., the client's and the therapist's. In a similar way, in the family therapy session, the voices the family members call upon while speaking may refer to real persons who are present or they may also refer to absent persons, even to fictional or imaginary characters. Although there is only one speaker telling the story, the storytelling evokes different voices in a dynamic interaction with each other. In contrast to earlier forms of systemic therapy, we do not just focus on the behavioral interaction among people in the room, but also look at the way the "inner voices" mediate the social interchange. That is why, in our dialogical analyses, we focus on voices, not on persons.

The richness of the family therapy conversation has to do not only with the polyphony of outer voices but the polyphony of inner voices. The latter becomes evident if we focus on those voices, which are not "seen" but sensed as present in each person's inner dialogues (Seikkula, 2008). While the therapist does not have access to the inner voices of the clients, they have knowledge of their own "inner speech." The therapist's own inner voices can refer to their professional self or personal one (Rober, 1999). Therapists participate in the dialogue in the voices of their professional expertise, e.g., as a doctor, a psychologist, having training as a family therapist, and so forth. In addition to such professional voices, the therapists participate in the dialogue in their more personal, intimate voices. If a therapist, for instance, has experienced the loss of someone loved or near, the voices of loss and sadness become part of the polyphony. This does not mean that the therapist would necessarily speak about their personal or intimate experience of death, but the voice of loss would affect how the therapist adapts to the present moment. The therapists' inner voices containing their own personal and intimate experiences become a powerful part of the joint dance of the dialogue (Seikkula, 2008), involving aspects such as how the therapists sit, how they look at the other speakers, how they change their intonation, at which point they break in, and so on.

These inner voices, evoked in the session, will not merely be present in the story; they will also become alive and part of the present moment. The inner dialogue between the voices of our lives is not so much a matter of focusing on the meaning of the words uttered, but of sensing the nuances of the present moment. As Vygotsky (1962), referring to the peculiarities of inner speech, noted, "the first [of these peculiarities] is the preponderance of the sense of a word over its meaning" (p. 146). What is more important than the "stable" meaning of the words said is the sense of the words in the actual present moment and context. The same words can generate very different types of meanings when used in different conversations, even if the same people are present.

Positioning

As the concept of voice is the central to analyzing polyphonic dialogues, the action of positioning gives complementary information about the dialogical process. For example, a person can speak in voice of a partner but from the varying tone or position of, i.e., being a victim as opposed to an agent. In systemic family therapy, the process of inquiry often includes positioning questions that invite a shift in position from passenger to a more active agent or from an unreflective to a reflective and reflexive stance.

Wortham (2001) distinguishes between *representational positioning*, referring to the positions of the subject in the story and *interactional positioning*, referring to the speaker, the addressee, and the audience in the present conversation. In family therapy, interactional positioning—how the family members position themselves in the present moment of the session—is highlighted. Positioning is usually not a unilateral act by the speaker aimed at manipulating others strategically in the conversation. Rather, positioning happens between people in the process of continuous responses to what is uttered.

In multi-actor dialogue, interactional positioning emphasizes as a starting point looking at who is taking the initiative, both regarding the content and the process of speaking (Linell, 1998). In an optimal multi-actor dialogue, there would be flexibility in terms of who is taking initiative, instead of the therapist consistently doing so. In a session, as family members contrast their perspectives with the positions that they attribute to others—as well as with the positions they are invited into by other interlocutors—the phenomena of conflict and disagreement arise that are important to notice. They refer to the continuous dance of the changing positions in the session, giving the therapists some sense of what is at stake for the family members. At the same time, the therapists are also invited to take positions in the family's performance.

Addressees

Every utterance has both an author and the person to whom it is addressed, as every utterance is a response to what has previously been said (Bakhtin, 1986). The utterance may be addressed to someone who is present in the same room. We can state our opinion directly about the issues under scrutiny; we can agree with what was previously said, we can object to it, we can partly agree, adding our own point of view to what has been said, and so on. In multi-actor dialogues, we typically speak to one person, but at the same time, we are very aware of those others who are present, and our speaking is modified because of their presence. In this sense, those others who are present are part of the addressed audience and become part of the utterances.

Bakhtin (1986) calls these people the addressees. But there is more, as in dialogue, a third party is always present, even if only two persons are speaking to each other. They are speaking in the present moment. And, at the same time, when they are speaking about emotional issues, they may be addressing their words to those nearest to them: their mother, father, or loved one. Bakhtin also speaks of the super-addressee present when we address our words to some ideology relevant to our life.

In analyzing dialogues, it is not always easy to recognize the addressees. If the addressee is defined as a person sitting in the same room, no difficulties emerge. The addressees may be referred to openly and can thus be defined. Or, they may be present in the speaker's inner dialogue, affecting his or her intonation, choice of words, body gestures, and many other things—but without being openly recognizable.

Conceiving therapy in this dialogical way poses research challenges. The most important of these challenges is to find ways to examine the dialogical qualities of therapeutic conversations. A number of research methods have been developed that aim at a deeper understanding of the dialogical qualities of therapeutic conversations (e.g., Leiman, 2004; Salgado, Cunha, & Bento, 2013; Stiles et al., 2004; Wortham, 2001). While we have been inspired by these methods (Leiman, 2004; Stiles et al., 2004) they all focus on dyadic dialogues, i.e., the dialogue between one therapist and one client. In this book, we are developing methods for the study of multi-actor dialogues, such as occur in couple and family therapy, network meetings, and group discussions. There is no literature on dialogical research on multi-actor meetings, except for the work of Markova, Linell, Grossen, and Orvig (2007) on the dialogical analysis of focus group interviews.

We have been especially interested in the responses generated in family therapy dialogues, both in terms of the outer conversation among family members and in the inner speech of each participant in the session (Olson et al., 2011; Seikkula, Laitila, & Rober, 2011) as well as the inner voices of each participant (Laitila, 2009; Rober, 2005a; Rober, Seikkula, & Laitila, 2010).

Our Research Approach: Exploring Response Processes

To start, the multi-actor session has to be video recorded and transcribed. Depending on the focus of the study, we transcribe either a specific part of the session or the whole session. To make a multi-actor perspective possible, the transcript of the therapy conversation is printed in columns, one column for each speaker. Utterances are written in the columns in temporal order (Table 5.1). For a successful exploration, one has to be able to read the text simultaneously with a video or audio recording of the session.

The research process proceeds in steps, as follows.

STEP I: Exploring topical episodes in the dialogue

Defined topical episodes are the main unit of analysis (Linell, 1998). Topical episodes are defined in retrospect, after the entire dialogue of one session has been divided into sequences. Episodes are defined by the topic under discussion and

Table 5.1 Rating of dialogue in Victoria's and Alfonso's three therapy sessions

	Semantic dominance	Interactional dominance	Monologue vs. dialogue in episodes	Indicative vs. symbolic language area
First session				
Victoria	14	9		
Alfonso	3	0		
Therapists	3	11		
Total	20	20	16 vs. 4	14 vs. 6
Second session				
Victoria	8	5		
Alfonso	6	5		
Therapists	7	11		
Total	21	21	12 vs. 9	13 vs. 8
Third session				
Victoria	6	6		
Alfonso	5	2		
Therapists	9	12		
Total	20	20	9 vs. 11	8 vs. 12

are regarded as a new episode if the topic is changed. The researcher can choose, out of all themes, some specific important topics for further analysis. After dividing the session into topical episodes, within each episode, certain variables are identified, as specified below.

STEP II: Exploring the series of responses to the utterances

In each sequence, the way of responding is explored. Responses are often constructed within a series of utterances made by each participant in the actual dialogue. Within each topical episode, the responses to each utterance are registered to gain a picture of how each interlocutor participates in the creation of the joint experience in the conversation. A three-step process is followed. The meaning of the response becomes visible in the next utterance to the answering words. It can start with whatever utterance is regarded as the initiating utterance (IU). The answer given to this IU is categorized according to the following aspects.

1. *The participant who takes the initiative (i.e., who is dominant) in each of the following respects.*
   Quantitative dominance: this simply refers to who does most of the speaking within a sequence. *Semantic or topical dominance*: this refers to who is introducing new themes or new words at a certain moment in the conversation. This individual shapes most of the content of the discourse. *Interactional*

dominance: this refers to the influence of one participant over the communicative actions, initiatives, and responses within the sequence. It is possible that this individual could have more influence on other people than that exerted by those actually speaking (Linell, 1998; Linell, Gustavsson, & Juvonen, 1988). For instance, when a family therapist is inviting a new speaker to comment on what was previously said, he or she can be said to have interactional dominance. At the same time, someone who is very silent also can have interactional dominance by evoking solicitous responses from others. Rather than identifying the person who is dominant in the family session, the main focus of our research is to understand the shifting patterns of these three kinds of dominance.

2. *What is responded to?*

The speakers may respond to

- Their experience or emotion while speaking of the thing at this very moment (implicit knowing)
- What is said at this very moment
- Some previously mentioned topics in the session
- What or how it was spoken
- External things, outside this session
- Other issues (If so, what?)

These are not mutually exclusive categories. In a single utterance, many aspects of experience can be presented. The special form of answers in a situation in which the speaker introduces several topics is considered to form one utterance. We look at how the answer helps to open up a space for dialogues in the response to that answer.

3. *What is not responded to?*

What voices in the utterance (bearing in mind that a single utterance by a single participant can include many voices) are not included in the response of the next speaker?

4. *How is the utterance responded to?*

Monological dialogue refers to utterances that convey the speaker's own thoughts and ideas without being adapted to the prior utterance of someone else. One utterance rejects another one. Questions are presented in a form that presupposes a choice of one alternative. The next speaker answers the question, and in this sense, his or her utterance can be regarded as forming a dialogue; however, it is a closed dialogue. An example would be when the therapist asks for information about how the couple made the contact, and the couple answers with information about their actions leading up to participation in the therapy session. In *dialogical dialogue*, utterances are constructed to answer previous utterances and also to wait for an answer from utterances that follow. A new understanding is constructed between the interlocutors (Bakhtin, 1984; Luckman, 1990; Seikkula, 1995). This means that in his or her utterance, the speaker includes what was previously said and ends up with an open form of utterance, making it possible for the next speaker to join in what was said.

5. *How is the present moment, the implicit knowing of the dialogue, taken into account?* These are the emotional reactions noted that emerge while speaking of an emotional issue.

In looking at videos of dialogues including sequences of responses, all the issues of ways of being present in the meeting will be observed, including body gestures, gazes, and intonation. Often this includes (for example) observing tears or expressions of anxiety—aspects not seen when one merely reads the transcript. The present moment becomes visible also in the comments on the present situation (e.g., comments on the emotions felt concerning the issue under scrutiny).

STEP III: Exploring the processes of narration and the language area

This step can be conducted in two alternative ways:

1. Indicative versus symbolic meaning

   This distinction refers to whether the words used in the dialogue are always being used to refer to some factually existing thing or matter (indicative language) or whether the words are being used in a symbolic sense; in other words, whether they are referring to other words, often describing more intangible experiences of emotional life, rather than to an existing thing or concrete matter (Haarakangas, 1997; Seikkula, 1991, 2002; Vygotsky, 1981; Wertsch, 1985). Each utterance is categorized as belonging to one of these two alternatives.

2. Narrative process coding system

The preliminary development of this coding system was undertaken by Agnus, Levitt, and Hardtke (1999) within individual psychotherapy. Laitila, Aaltonen, Wahlström, and Agnus (2001) further developed the system for the family therapy setting. Three types of narrative processes are distinguished. The speaker uses (a) *external language*, giving a description of things that happened; (b) *internal language*, describing his or her own experiences of the things he or she describes; or (c) *reflective language*, exploring the multiple meanings of things, the emotions involved, and his or her own position in the matter.

Concerning the core idea of looking at multi-actor dialogues, it is preferable to work with a team of researchers. This is not only because it enhances the credibility of the research, but also because team investigations into multi-actor dialogues seem appropriate when one sees analysis as a multi-actor process. One possible way is to start by having a single researcher analyze the transcript. After the preliminary categorization (using the three steps described previously), the research team comes together to review the video of the session and also the transcript. In the course of that meeting, as a check on the trustworthiness of the first author's analysis, the co-researchers review the categorization, focusing more on their points of disagreement than on their points of agreement. In dialogue with each other, the different voices enrich the picture of the dialogue in focus.

STEP IV: Microanalysis of specified topical episodes

After the classification of the above-mentioned variables, a new step in the process
of investigation is taken by choosing specified topical episodes as subjects for
microanalysis. In the first phase, an answer was received what happens within
the topical episodes in general and in comparison to other topical episodes and
thus to see the points, in which the change has started to happen. We will be looking in detail what happened in the specific events of change. In this analysis, the
above-mentioned concepts of voices, positioning, and addressees can be used,
but the microanalysis can be conducted in other ways also. In the tables it will be
illustrated, how these concepts can help us to pick up essential elements of what
happened.

Development of Dialogue in the Therapy Sessions of Alfonso and Victoria

Each session was investigated by the DIHC method. First, the topical episodes
were defined. In the following description only the three first sessions are registered, because the fourth one was a shorter evaluation session of how they felt
at the time of the interview and how they evaluated the previous therapy
sessions.

As seen in the summary presented in the Table 5.1, the quality of the dialogue
varied from session to session. In the first session, Victoria mostly took both semantic and interactional initiative. Alfonso mainly responded to the themes that Victoria
raised. In only three instances did he take the initiative regarding the subject. This
was quite different in the second and third sessions. Alfonso became as active as
Victoria in terms of initiating the subject of the conversation and also in asserting
interactional dominance, challenging Victoria regarding how a particular issue is
talked about. Therapists' activity increased in such a way that in the third session, it
was therapists who initiated most of what was talked about. Out of altogether 61
topical episodes during the three first sessions, 18 focused on their relation to
Alfonso's family.

Concerning the emergence of the dialogical quality of the conversation, it was
the third session that was different from the first two. Most of the topical episodes
were dialogical ones. The same kind of change happened in relation to the language area. While most of the topical episodes during the first two sessions used
indicative language, the third one showed many more exchanges of symbolic
meaning. These statistics illustrate well that in the third session the couple
reflected on a lot of things having to do with their relationship with each other and
with Alfonso's family and the impact of the cultural differences in their family
backgrounds.

Looking at Voices, Addressees and Positioning in the Micro-analysis

Because the most notable change seemed to happen during the second session, in the following a detailed inquiry of three specific episodes in the second session will be given. The following three episodes illustrate the shift that took place in the quality of dialogue during the meeting.

1. *Proceeding in dialogue in the beginning of the second session*

The beginning of the session seemed rather chaotic and tense. When Therapist 1 entered the waiting area, he saw Victoria and Alfonso coming into the building. There was some additional commotion, and Alfonso had to convince Victoria to come into the office.

In the beginning of the session, the therapist asked the couple how they had been feeling and Alfonso said, "quite good." Victoria responded feeling differently and not wanting to come to the session. The therapist answered her utterance in his further question, in which he implicitly chose to respond to the specific concern most active in the present moment, which is an invitation to be in dialogue. From the dialogic viewpoint, her expression of "tiredness," may have embodied the perceived absence of the dialogic context: namely, a new, joint language in which she could give voice to her real feelings. Her comment was monological in the sense that she was declaring the way things are, and not making room for the contributions of a listener.

Victoria started to talk about her relationship with Alfonso and spoke in the voice of a forlorn and disappointed partner, twice repeating the word "sad." Victoria made a negative comment about a homework assignment suggested by the therapist. She conveyed implicitly that the suggestion made her feel defeated, because her partner did not do the homework.

In this exchange, Alfonso's voice merged with hers. He incoherently tried to back up her statements, but he became virtually incomprehensible. Alfonso's replies consisted of sentences he started but did not finish. It was as if Alfonso created a mist of words in which nothing was really stated outright.

After Alfonso's talk, Victoria expressed a depressing thought about the effect of her work: "I've been working like approximately 15 hours per day and I am never home, and then if I am home, he is not because he has some friends to see or something..." She remarked that she was never home at the same time as Alfonso.

The therapist responded by saying:

T1 You said never, what does it mean?

V That I am never home?

T1 Not that you said that you are never home, that you are never home together?

This question of Therapist 1 about the meaning of "never" seemed to mark the shift toward Victoria becoming more direct and explicit about her concerns. The therapist's question was a successful attempt at generating a dialogue with her, although it came in the form of a subtle and tense challenge to Victoria's hopelessness. The clear voice of the therapist himself appeared to make the difference by showing both that he was intimately listening by his repetition of her words while simultaneously resisting her despairing conclusion.

By this simple question Victoria's utterance was brought in relation to another voice. A dialogue was born; the response gained Victoria a voice, and she was no longer alone. Alfonso also changed after that and suddenly began to express himself clearly in an assertive voice that also dissented from that of his girlfriend. This small segment illustrates how the therapists' minimal response in dialogue helped to reconstitute the entire context as a dialogic one and to set in motion remarkable changes in the interchange between the couple.[1]

2. *Change in dialogue about the relation to Alfonso's family*

As Alfonso became more lucid in the session, the therapist asked how the couple wanted to use the meeting. In response, the couple agreed to discuss the partner's upcoming trip to his home country. This topic led into talking about the most difficult and complex issue in the relationship, the relationship of the couple to Alfonso's family. Two years earlier, Victoria had visited Alfonso with his family at the family home. While there, she tried to make a connection with Alfonso's mother, but his mother denied her overtures and refused to say anything to Victoria. It was after this visit that Victoria became depressed to the point of qualifying for disability, which she was on for 1 year. The discussion of this topic will be analyzed in two episodes that illustrate a change in the way the couple handles the issue of their relationship to Alfonso's family (Table 5.2).

3. *Confusion in utterances in the dialogue*

The episode opened by Alfonso describing his visit home, which Victoria was not part of. Victoria responded to his opening by taking offense, to which Alfonso responded by defending himself. Victoria's next comment took back the offended tone, thus repositioning herself as more agreeable. The entire episode stayed monological. They addressed utterances to each other in the room, but for Alfonso, his family seemed to be a difficult subject to discuss, while, for Victoria, especially harmful because of the traumatic experiences with Alfonso's mother. The halting nature of this conversation suggested the influence of the prior experiences, or inner voices of loyalty and trauma, respectively, which remained invisible. Even when the therapist tried to comment on the emotions implicit in the present moment, no dialogical change occurred that reduced the difficulty of the interchange.

---

[1] Part of this excerpt is described in Olson, Laitila, Rober, and Seikkula (2012).

Table 5.2 Dialogue sequence 16′ from the start in the second session

Transcript session				Categories	Comments
Victoria	Alfonso	Therapist 1	Therapist 2		
	Well I am going to (A's home country)for one week, so…				New episode: visit to home
		yhm…			
	.. I think there will be like good time to see… For example when I went the last time, these things that…like…you would like to me to send messages it would be like good to see…			A semantic dominance Voice of last visit Address to V	A in between V and home connections
		What are your thoughts about it, about the visit?		Answer to the "thoughts of the visit" not the uncertainty in A's utterance	
	(7) maybe I will like to think that even if some days particularly… maybe I can't or then happens that I can't.. even if we talk that you could try.. to feel it back' …			Responds to V, not the therapist Voice of people at home	
What do you think you couldn't … send me a message a day?				V interactional dominance Positioning as needing contact and offending	

		Voice of the people at home	Discussion seems not to develop into dialogical
I only told you that I can send you like I told you and there are like a week, if there is during this week one day, even if I call, maybe it can be or maybe not or it can be that I am really busy… I can do this, I can send you message but if for one day it will happen once that maybe I can't send you a message, once. Like maybe if you would try to understand that it could happen…		Positioning as asking for understanding from beneath; defending his point of view	
Yeah, I try. I just think that it is quite easy thing to do, if you want to		Answers both to the request and to the dilemma between her and A's people at home	
yes, but I can tell you that sometimes it can happen that if in any way to call you in the evening, in some time it can happen, but that… if I have many many things to do it can happen that if we call in the evening…		Voice of the people at home — Address to the voice of home	Monological utterances
No.. that's for sure that we can't call every evening		Positioning herself as the one not offending	
Why?			
Because it has been so always. I don't even need that I don't even need that..		Positioning as sacrificing herself	
okei (7)			
			New episode: V being afraid

(continued)

Table 5.2 (continued)

Transcript session

Victoria	Alfonso	Therapist 1	Therapist 2	Categories	Comments
		And V what are your thoughts about…		T1 interactional dominance	
[well I am a afraid because]				V semantic dominance Positioning as being afraid	
		Alfonso's visit to his home country?			
Now he has been there for three times alone and every time I have got sad sometimes because I feel that he really doesn't have time to send me a text message that how can it really be like… I just feel like I am nothing to him when he is there.. sometimes…				Voice of last visits— being sad. Positioning as week and worthless. Address to A and to the people at A's home	Key experience concerning becoming depressed?
		yhm..			
But every time that he has been at some point or many times I have felt like that…					
		yhm.. felt like?		Answers to the emotion	Repeating A try to dialogical dialogue?

That he doesn't think about me and he kind of likes to forget about me when he is there.			Positioning as being poorer than A's people at home
	And what does it mean that you think… (Shows by her hand to A) he would like forget you?		Answer by asking meaning — being at the present moment
I feel like, I don't know… It don't make any sense to me, how can A be so busy that he doesn't have time to think about me or send a message to me.. that just something that I don't understand. I don't need a phone call. I just need to know that you think about me			Voice of being abandoned

Positioning between A's home people and her self

Aiming at "forcing" A to choose? |
| But I think it has never been.. it has never been that for one day I didn't send a message to you or phone call | | | Does not answer to her feeling |

Alfonso Finds Words to His Position Between Victoria and His Family

This episode starts with Alfonso's attempt to describe his position while visiting home. Therapist clarifies Alfonso's statement by repeating word by word what he said: "I didn't quite follow what you said, Alfonso, that every time you have been separated so you have…" Alfonso answers the therapist. Perhaps it is easier for him to speak to the therapist instead of speaking directly to Victoria. At least his utterances are now much more complete and clear. When the therapist clarifies this by saying that "it seems to have to come a big issue in your relationship," Victoria responds by positioning herself no longer as attacking Alfonso, but instead of protecting him. She becomes more understanding and says: "I don't want Alfonso to be between two families…" This opens up a sequence, in which the relationship between Alfonso and his family and Victoria is clarified, defining it in a more mutual and detailed way (Table 5.3).

The dialogue between the couple, in which they began directly addressing each other, inhabited the main part of the end of the second session. For the first time, each spoke from a distinct "I" position, addressing their partner as "you." Victoria and Alfonso took the risk of confronting each other. Both therapists were more on the outside now. At this point in the session, there were 20 short exchanges between the couple, without any question or remark from one of the therapists. For example, Alfonso recounted a telephone call he made to talk with her when she was taking the bus home from work. Victoria agreed that that made her happy. They went on to discuss how Victoria wanted further signs of commitment, while Alfonso wanted to go out with his friends without having to fight about it with her.

The essential change toward a dialogical equality happened in this second session. Participants' different voices were present and heard. The change actually seemed to happen in the above-mentioned episodes and every participant in the session contributed to it. Both Alfonso and Victoria spoke to each other in quite a different manner compared to when they started. Both therapists were actively responsive especially to those particular utterances that having to do with what has happening in the here-and-now present moment. What was alive in the dialogue in this way were the sensitive issues of the relationship with Alfonso's family that earlier seemed to have been related to Victoria becoming depressed. In this sense, the inner voices of the dialogue and how they mediated their ongoing interaction became more openly expressed and reconfigured. One consequence of this shift was an overall repositioning of each partner from feeling that they were an emotional victim of the other to becoming agents in the joint negotiation of their relationship.

Table 5.3 Dialogue sequence 42′ from the start in the second session

Transcript session

Victoria	Alfonso	Therapist 1	Therapist 2	Categories	Comments
	1. I am just thinking that it always that if we have been together to any place that I've been away that you haven't. When I am there for you its more, its not to kind of normal time like normal day, it…			Semantic dominance Monological Positioning defending himself	
		I didn't quite follow what you said A that every time you have been separated so you have…		Dialogue	Tries to intervene in a dialogical way
	Like I've always been, she has been here and I was like to (A's home country)				
	…so to me its like different.. day, different…	okei			
	yes, its not the same situation, we are in two different places. It is.. we are apart… still I think, I mean different context.	[so its not the same..]		Having more words for the complicated relationships	Speaking at the same time

(continued)

Table 5.3 (continued)

Transcript session

Victoria	Alfonso	Therapist 1	Therapist 2	Categories	Comments
		It seems to bit… It seems to have to come as big issue this and in a very concrete detail in your relationship…			The balance between doing and feeling, V speaks of her emotion in practical terms whereas A responds to concrete actions and T1 is trying to combine these two perspectives. T needs to speak with two different "second voices" or "hidden" voices. V has a hidden voice, A don't catch the point
yeah.. and it is a big thing and I know that family is important and I also I have tried also to… I don't want A to be between two families…				Dialogical / Responds to A / Address both A and T / Understanding / Decreasing her expectations	
…but I think I have tried then I feel that do they want to keep so busy so that you don't have time for me, because it's clear that they don't like me.…		[yhm]	yhm..		Combines with her emotional experiences of being less important than A's family / First time hidden voice into an open voice

		Which kind of situation.	Dialogical
That he has to be between two fires..			Responds to "situation": the complicated relationships entity
	aaa.... you think that A's family likes him to be in between…		Metaphorical, powerful description
			Dialogical
			At the present: "You think…"
Yes it seems like, because I have … like we have tried…			Dialogical
	yhm…		
Yeah. like there's also this difference that I just don't care what my family would tell me. Or if they don't come here to visit, it really is not important. If it would be just for me…			Responds in one part to V
			Positioning still as defending himself
			Address to his family
	okei..		Is A convincing?
And then what she tells me that it would be more like…			
	yhm…		

(continued)

Table 5.3 (continued)

Transcript session

Victoria	Alfonso	Therapist 1	Therapist 2	Categories	Comments
	that if it would be a kind of important for them…				
		yhm…yhm…			
	…my life here they would come here to visit.				
I don't think its fair for A that he has to be in the situation that he has two families and he has to like… That he is the one who travels to (…) all the time.				Dialogical Respond to A's position Is present and deliver understanding	
	To me it's not a big deal because any way when I go home it's not just for them, it's also like friends and everything…			Dialogical Adding friends	Less emotionally loaded issue
		yhm… yhm…			

Discussion and Conclusion

Concerning our second aim for this chapter, we noticed that employing Bakhtinian concepts as research tools, there were specific moments in the session that were noted in which the positive change started to occur. In these moments, the therapists assisted in new ways of speaking by their responses to the utterances of the couple. The core problem that the dialogue addressed was the complex relationship of the couple to the Alfonso's family, especially to his mother. While answering first, the sad, despairing words of Victoria and then, the confusing expressions of the partner, the therapists assisted in the emergence of alternate voices. The shift in the voice of Victoria from anger and sadness to understanding and that of Alfonso from confusion to coherence allowed for the possibility of a joint discussion. New decisions and a different, concrete direction occurred regarding Alfonso's next visit to his family of origin. This change allowed Victoria to experience a greater sense of emotional commitment from her partner, thus making her happier.

Dialogical Investigations of the Happenings of Change was used in the same case earlier, when we examined one topical sequence of the very first minutes in the second session (Olson, 2012). By analyzing additional sequences, we have been reinforced in our earlier conclusion that the spare responsiveness of the therapist can be quite effective. The therapists' contributions are highlighted in specific episodes and assisted the changes that occurred from the couple's monological way of being to a more dialogical way of being in dialogue. In the latter kind of exchanges, the participants revealed more of a sense of agency in relation to their difficult, life-important dilemma.

Concerning our first aim, we noticed that the DIHC method allows for a detailed look at what happens in the specific moments of change. In the investigation of the couple therapy of Alfonso and Victoria, we found that there was a real development that happened from session to session in terms of the ways the partners participated in and co-created the dialogue. Each became more of a "subject" in their utterances during the discussion of the critical issues, and thus, more of a subject in their lives.

Finally, our experience from using the method *Dialogical Investigations of the Happenings of Change* we have found that it helps to synthesize large amounts of information about couple therapy and see both the core element of dialogical exchange and their variation from session to session. This research approach makes it possible to handle complex data in qualitative research of family therapy dialogues by helping to identify, distill, and analyze the transformative sequences. Dialogical investigations are a new addition to discursive analytic methods. Our hope is by this example to show its specific way of helping to see the flow of dialogue and how this flow of dialogue is related to the conduct of the therapist who helps create a dialogical context for the specific events of change. By looking at the dialogue in a couple therapy session from the beginning to the end, we can see the entire process as a flow of utterances and development.

References

Agnus, L., Levitt, H., & Hardtke, K. (1999). The narrative process coding system: Research applications and applications for therapy. *Journal of Clinical Psychology, 50,* 1244–1270.

Bakhtin, M. (1981). *The dialogic imagination.* Austin, TX: University of Texas Press.

Bakhtin, M. (1984). *Problems of Dostoevsky's poetics.* Minneapolis, MN: University of Minnesota Press.

Bakhtin, M. (1986). *Speech genres and other late essays.* Austin, TX: University of Texas Press.

Haarakangas, K. (1997). The voices in treatment meeting: A dialogical analysis of the treatment meeting conversations in family-centred psychiatric treatment process in regard to the team activity. Diss. English Summary. *Jyväskylä Studies in Education. Psychology and Social Research, 130,* 119–126.

Laitila, A. (2009). The expertise question revisited: Horizontal and vertical expertise. *Contemporary Family Therapy, 31,* 239–250.

Laitila, A., Aaltonen, J., Wahlström, J., & Agnus, L. (2001). Narrative process coding system in marital and family therapy: An intensive case analysis of the formation of a therapeutic system. *Contemporary Family Therapy, 23,* 309–322.

Leiman, M. (2004). Dialogical sequence analysis. In H. Hermans & C. Dimaggio (Eds.), *The dialogical self in psychotherapy* (pp. 255–270). Sussex: Brunner-Routledge.

Linell, P. (1998). *Approaching dialogue: Talk, interaction and contexts in dialogical perspectives.* Amsterdam/Philadelphia, PA: John Benjamins.

Linell, P., Gustavsson, L., & Juvonen, P. (1988). Interactional dominance in dyadic communication: A presentation of initiative-response analysis. *Linguistics, 26,* 415–442.

Luckman, T. (1990). Social communication, dialogue and conversation. In I. Markova & K. Foppa (Eds.), *The dynamics of dialogue* (pp. 45–61). London: Harvester.

Markova, I., Linell, P., Grossen, M., & Orvig, A. S. (2007). *Dialogue in focus groups: Exploring socially shared knowledge.* London: Equinox.

Olson, M., Laitila, A., Rober, P. & Seikkula, J. (2012). The shift from monologue to dialogue in a couple therapy session: Dialogical investigation of change from the therapists's point of view. Family Process, 51, 420–435.

Rober, P. (1999). The therapist's inner conversation: Some ideas about the self of the therapist, therapeutic impasse and the process of reflection. *Family Process, 38,* 209–228.

Rober, P. (2005a). The therapist's self in dialogical family therapy: Some ideas about not-knowing and the therapist's inner conversation. *Family Process, 44,* 477–495.

Rober, P. (2005b). Family therapy as a dialogue of living persons. *Journal of Marital and Family Therapy, 31,* 385–397.

Rober, P., Seikkula, J. & Laitila, A. (2010). Dialogical analysis of storytelling in the family therapeutic encounter. Human Systems Journal, 21, 27–49.

Salgado, J., Cunha, C., & Bento, T. (2013). Positioning microanalysis: Studying the self through the exploration of dialogical processes. *Integrative Psychological and Behavioral Science, 47*(3), 325–353.

Seikkula, J. (1991). *Perheen ja sairaalan rajasysteemi* [Family-hospital boundary system in the social network. English summary]. Jyväskylä studies in education, psychology and social research, 80.

Seikkula, J. (1995). From monologue to dialogue in consultation with larger systems. *Human Systems, 6,* 21–42.

Seikkula, J. (2002). Open dialogues with good and poor outcomes for psychotic crises: Examples from families with violence. *Journal of Marital and Family Therapy, 28,* 263–274.

Seikkula, J. (2008). Inner and outer voices in the present moment of family and network therapy. *Journal of Family Therapy, 30*(4), 478–491.

Seikkula, J., Laitila, A., & Rober, P. (2011). Making sense of multifactor dialogues. Journal of Marital and Family Therapy, 37. doi: 10.1111/j.1752-0606.2011.00238.x.

Seikkula, J., & Arnkil, T. (2014). *Open dialogues and anticipations: Respecting otherness in the present moment.* Helsinki: Thl.

Steiner, G. (1989). *Real presences: Is there anything in what we say*. Chicago: University of Chicago Press.

Stiles, B., Osatuke, K., Click, M., & MacKay, H. (2004). Encounters between internal voices generate emotion: An elaboration of the assimilation model. In H. Hermans & C. Dimaggio (Eds.), *The dialogical self in psychotherapy* (pp. 91–107). New York: Brunner/Routledge.

Vygotsky, L. (1962). *Thought and language*. Cambridge, MA: MIT Press.

Vygotsky, L. (1981). The development of higher forms of attention in childhood. In J. Wertsch (Ed.), *The concept of activity in Soviet psychology* (pp. 189–240). New York: M. E. Sharpe.

Wertsch, J. (1985). *Vygotsky and social formation of mind*. Cambridge, MA: Harvard University Press.

Wertsch, J. V. (1991). *Voices of the mind: A sociocultural approach to mediated action*. Cambridge, MA: Harvard University Press.

Wortham, A. (2001). *Narratives in action: A strategy for research and analysis*. New York: Teachers College Press.

Chapter 6
Fostering Dialogue: Exploring the Therapists' Discursive Contributions in a Couple Therapy

Evrinomy Avdi

Several authors have argued convincingly that, despite the extensive literature that exists on theories of psychotherapy and change, we know relatively little about how change actually happens in psychotherapy; this situation renders qualitative, discovery-oriented research on therapy sessions particularly useful (McLeod, 2011). The present chapter describes such a qualitative study, which aims to illuminate the interactional and discursive processes that underlie therapeutic work that was carried out with a couple from a dialogical perspective. The questions that guided the analysis concerned, firstly, the shifts that could be observed through the sessions in meaning construction, in positioning and in interaction and, secondly, the therapists' contributions to these shifts.

Narrative, Discourse and Dialogue

This study is located in the growing contemporary literature that approaches psychotherapy in terms of discourse, narrative and dialogue, a trend that has been influential in both individual and family therapy. Although these approaches are diverse, they share a view of therapy as a process entailing the reconstruction of meaning and the reformulation of the clients' subjectivity. In this context, it is assumed that psychological problems constitute discursive and interactional phenomena, which are created, maintained and dissolved in and through language and interaction. Therapy is accordingly conceptualised as providing a relational and conversational context in which clients reconstruct their life narratives in ways that are increasingly complex, evocative, inclusive, polyphonic and flexible (Anderson, 2012;

E. Avdi (✉)
School of Psychology, Aristotle University of Thessaloniki, Thessaloniki, Greece
e-mail: avdie@psy.auth.gr

© Springer International Publishing Switzerland 2016
M. Borcsa, P. Rober (eds.), *Research Perspectives in Couple Therapy*,
DOI 10.1007/978-3-319-23306-2_6

Anderson & Goolishian, 1988; Angus & McLeod, 2004; Hermans & Dimaggio, 2004; Seikkula, 2011; Seikkula & Arnkil, 2006; White & Epston, 1990).

Adopting a broadly narrative view on psychotherapy calls for research approaches that allow us to study the process of therapy in terms that are consistent with the underlying theory and epistemology. However, the majority of qualitative research on psychotherapy relies on grounded theory or IPA, whereas the potential of herme-neutic, constructivist and social constructionist epistemologies has not yet been fully tapped (Elliott, 2012; McLeod, 2011). This is arguably particularly pertinent in the case of family and couple therapy given their epistemological allegiance to constructionist, narrative and dialogical principles.

In this chapter, I argue that discourse analysis is commensurate with the narrative and constructionist underpinnings of couple therapy and that it provides useful tools and concepts for studying psychotherapeutic dialogues. Moreover, in line with others (e.g. Strong, Busch, & Couture, 2008) I propose that different conversational aspects can constitute valid and useful 'outcome variables' in therapy process research and so studying language-in use (rather than relying on retrospective accounts or coding therapy talk in terms of predetermined variables) can be a useful focus for process-outcome research. In this study, several inter-related conversational aspects were examined, including meaning construction, discourse use, subject positioning, blame management and the rhetorical strategies used by participants.

Examining Meaning Construction and Interaction in Therapy

Psychotherapy is an interactional situation that calls for—indeed relies upon—the telling, exploration, reformulation and resolution of clients' 'problems'. As several discursive studies have shown, an important part of the work of therapy is carried out through transforming the clients' complaints into 'problems' that can be under-stood within a psychotherapeutic frame and resolved through therapy (e.g. Buttny, 2004; Davis, 1986; Guilfoyle, 2001; Hak & de Boer, 1995). Moreover, talking about problems implicates issues of accountability, responsibility and morality, and so studying problem constructions in therapy also entails examining how speakers position themselves and others vis-à-vis the problem, primarily with regards to responsibility, accountability and agency. Furthermore, analytic emphasis of much qualitative research on psychotherapy is often placed on the clients' talk, whilst the therapists' discursive contributions often remain un-analysed or represented as 'simply' facilitating the reconstruction of client narratives (e.g. Avdi & Georgaca, 2007a, 2007b). This selective focus does not do justice to the interactional nature of psychotherapy and tends to conceal its ideological dimensions as well as its power. The need to examine interaction in its context is also supported by the findings of process-outcome research, which studies the relationship between specific process variables with therapy outcome. In contrast to the rationale of much quantitative process-outcome research, it seems that most commonly used interventions are not correlated with outcome (Lambert & Ogles, 2004; Stiles, Honos-Webb, & Surko,

1998; Stiles & Shapiro, 1994). Moreover, those variables that are consistently found to be correlated with outcome (e.g. therapeutic alliance, empathy, goal consensus; Norcross, 2011) tend to be the products of responsive action, rather than process variables per se (Barkham, Stiles, Lambert, & Mellor-Clark, 2010). Similarly, the findings from common factors research in both individual (Wampold, 2001) and family and couple therapy (Sprenkle, Davis, & Lebow, 2009) point to the central role of relational factors, and primarily the therapeutic alliance, to therapy outcome. Based on the above, it is suggested that the ways in which problems and identities are constructed and reformulated within the therapeutic interaction can form an important focus for therapy process research.

The material used for the purposes of this study was drawn from a couple therapy with a highly experienced family therapist who adopted a dialogical perspective. Dialogical approaches to therapy do not constitute a unitary or coherent school and several approaches have been articulated within this trend (Anderson, 2012; Bertrando, 2007; Hermans & Dimaggio 2004; Salvatore & Gennaro, 2012; Seikkula, 2011; Seikkula & Arnkil, 2006). In brief, therapy is broadly conceptualised in terms of dialogical processes, which are thought to take place both within participants, in the form of ongoing internal dialogues between different voices of the self, and between them. Such internal and external dialogues take place simultaneously and constitute the dynamics that arise in the therapeutic encounter. A shared assumption in several dialogical approaches is that a particular kind of dialogical conversation is potentially healing in and of itself, as dialogue permits the generation of not-yet said meanings (Anderson, 2012). Within the context of a dialogical encounter, space is created for emotional expression, the articulation of new meanings, the emergence of silenced or disowned voices and the reorganisation of the clients' position repertoire. In this literature, great importance is placed upon the therapists' dialogical stance, associated with a particular way of listening and responding to the clients' utterances. Related key concepts include the adoption of a 'not-knowing' stance, joint action and the recognition that meaning is unfinalisable (Guilfoyle, 2003); furthermore, the therapists' comprehensive, responsive participation in the present moment is considered crucial for the creation of dialogue (Seikkula, 2011; Seikkula & Olson, 2003; Seikkula & Trimble, 2005).

In sum, this study is located within a qualitative and constructionist research tradition; it follows a practice-based naturalistic design and aims to examine in detail how dialogical couple therapy 'gets done'. The focus of the analysis concerns the negotiation of the meanings implicated in constructing the problem and the clients' subjectivity, as well as the therapists' discursive contribution to this process.

Method of Analysis

The material for this study consisted of the audio-recordings and verbatim transcripts of four sessions of a successful couple therapy with a young, cross-cultural couple (for a description of the case see Chap. 2) and discourse analysis was used.

Discourse analysis (DA) is a broad and diverse field that includes a variety of different approaches to the study of language-in-use, which share a view on language as constructive, functional and variable. DA studies tend to examine the discursive processes through which reality, agency and accountability are negotiated within interaction (Wetherell, Taylor, & Yates, 2001).

In brief, in DA language is approached as a means of constructing—rather than mirroring—reality and as a form of social action, in the sense that speakers use language in order to achieve certain interpersonal goals. Broadly speaking, discourse analyses tend to examine *how* certain issues are constructed in talk, noting the *variability* in speakers' accounts, and explore the *rhetorical* aspects and the *functions* of talk in the context of the ongoing interaction. With regards to the process of analysis, different trends exist; for the purposes of this chapter, I will introduce key notions and tools for analysing discourse use in the context of therapy, with a leaning towards Foucauldian discourse analysis (Georgaca & Avdi, 2012).

Following detailed verbatim transcription and several close readings, open coding is initially performed on the text; this allows immersion in the material and leads to a fine-tuning of the research question and the selection of a corpus of extracts deemed relevant. Analysis is a systematic and interpretative non-linear process that involves several iterative cycles between different levels, which are described below.

Given that talk is approached as constructing the objects to which it refers, the initial focus of analysis is usually on *the construction and negotiation of meaning*. This entails an examination of the various ways in which the issue under study (e.g. the 'problem') is constructed through the sessions. In practice, this means that we examine all instances in the text where the object of study is mentioned or implied and we note both the regularities and the variability in these constructions. Then, we broaden our focus to locate these constructions within specific *discourses*, i.e. systems of meaning that are related to the interactional and wider sociocultural context and that operate regardless of the speakers' intentions. This is a step towards linking the specific interaction with ideology, institutions and power/knowledge.

A second level of analysis, which differentiates DA from several qualitative approaches that focus exclusively on the content of talk, concerns *the function of language*. Here, we examine the ways in which participants use language in order to achieve specific interpersonal goals, such as attributing responsibility, refuting blame or justifying one's actions. At this level of analysis, we examine how accounts are organised and the rhetorical strategies speakers use, in order to present their views as credible and themselves as objective, reliable and rational subjects. There are several ways in which the function of an utterance can be deduced, for example lexical choice, syntax, pauses, turn-taking, etc. Moreover, it is important to examine each utterance in the discursive context in which it was produced, by attending to sequence, i.e. what came before and what followed. Our general question at this level of analysis is: 'what is the function of talking about the issue in this way at this point in the interaction?' In addition to the microanalytic focus on interactional sequences, the function of talk can also be examined on a more macro-level, through studying, for example, a whole session. Relevant here is the notion of a speaker's *discursive agenda*; this refers to the effects a participant's talk has on the overall interaction. The agenda of each participant can be deduced only after the detailed analysis of the function of his/her talk, and is a very useful notion for analysing therapy transcripts.

Another important notion used in analysing discourse is that of *subject positioning*; subject positions refer to the identities made relevant through specific ways of talking. The notion of positioning emphasises the location of the person in discourse and within a moral order, and reflects the links proposed in post-structuralism between discourse and subjectivity (Davies & Harré, 1990; Harré & Van Lagenhove 1998; Henriques, Hollway, Urwin, Venn, & Walkerdine, 1998). In brief, discourses are assumed to entail an array of subject positions that people take up when they talk, often irrespective of their intentions and outside of awareness; these positions influence both the speakers' sense of self and the course of the interaction, as subject positions carry with them different rights and responsibilities. The main thrust of this argument is that self-descriptions, although may be experienced as authentic self-productions, actually reflect a 'selection from the panoply of selves already available to be donned' (Wetherell & Edley, 1999: 343) in a process that often remains unrecognised. In practice, we examine subject positioning in terms of (a) how one is positioned in particular interactions, i.e. the position from which one speaks and the position in which one places the person they address, and (b) how one is positioned through particular discourses, i.e. to the types of subject made available through the deployment of different discourses. The notion of subject positioning is very useful in examining the processes through which the client's subjectivity is constructed within the clinical dialogue.

Following intensive microanalysis as described above, another two levels of discourse analysis can be performed; these concern the effects of discourse use on practices and on subjectivity (for a description see Georgaca & Avdi, 2012). Due to space constraints and the research questions examined, these levels of analysis were not carried out in this case.

Findings

With regard to the process of analysis, the sessions were initially examined in relation to meaning construction; the analysis focussed on the different, and shifting, ways in which the 'problem' was constructed by the speakers, the discourses these diverse constructions drew upon and the subject positions associated with them. Next, questions regarding the function of constructing the problem in specific ways at specific points in the interaction were explored. Finally, the analytic focus turned to the question of how these changes came about, through studying closely the therapist's utterances and the effects they had on the ensuing discussion; based on this, the therapist's discursive agenda was formulated. Due to space constraints, the analysis of only two extracts is presented, although it must be borne in mind that all four sessions were analysed.

Initial Problem Construction

Reporting a problem involves representing an event or issue as perplexing, difficult and uncertain (Buttny, 2004); problem descriptions almost always implicate a cause, ascribe responsibility and blame and also imply solutions, both possible and ideal. A rather long stretch of talk from the very start of the first session is shown below, which illustrates the initial problem construction as well as the therapist's response to the clients' 'problem talk'.

Extract 1: constructing the problem and exploring solutions

Session 1, turns 7–8, 14–19, 50–60, 85–95

7T	OK (.) so where would you start?
8V	We can start from the beginning [giggles], it's, the situation is that, it was summer [2 years earlier] that I started, getting, I got depressed, and now now my situation is better (text omitted)) but the reason why we are here is that this, my thing left kind of scars in our relationship (…)
14A	Yeah, but you, yes, the thing it's just that [clears his throat] after this time now, that, that, if we can say that she's a lot better it means that she feels better now but still when we have this kind of situations that, earlier I felt I was aa I had kind of more patience, I could, I could go, like how'd you say (.) I had more patience in listening
15V	In comforting me
16A	Yes, yes, now I feel that it's like I don't, somehow even if there is some small thing
17V	Some small thing that I would like to discuss about, for example
18A	But then I think that I, somehow I react too strongly like if it would be something bigger
19T	So you are saying that meanwhile Victoria was down you could support her but now when things are better, in some way
50A	(…) I was like trying to support as much as, as I could so it wasn't, let's say a problem for me to deal with this kind of things, it was tough, of course, but then it it wasn't a problem
51T	What was it that made it tough?
52A	I'm sorry?
53T	What was the thing that made it tough? You said it was tough
54A	Ah, yes, yeah, of course, it was tough because it was like when she got like in the bad feeling that she didn't trust what I was saying, so it was for these two days I was always repeating things, trying to let her, to make her calm down
55V	But I I didn't believe anything he said, I just had this own imagination that was running so wild that no matter what anyone said but he, of course he had to, of course, still try
56A	That was the tough
57T	Yeah, that was the tough side that you in a way were forced to repeat things that was said and then
58A	Yes then somehow maybe now I feel a bit that, now that she has got better, so maybe now I feel that I don't, I've kind of used somehow maybe this like somehow that I've been, how'd you say, now I don't feel that I have this this perseverance, this, patience, as I used to have somehow
59T	Where it is disappeared?
60A	[Laughing] I don't know
85T	You could talk when you were down but now no longer? It's no longer possible

86V	Mmm (.) mmm
87T	And do you have any explanations or idea for yourselves to understand this change?
88V	I think it is how he said yes but now it has been like, we've been, waiting that this situation gets better by itself, but now it has been a half year and it it hasn't, so that's why I'm afraid that maybe it doesn't get better by itself
89T	And by, you said that the situation gets better what what, what does it mean the situation gets better?
90V	That Alfonso could talk with me, that when I need to talk about something he doesn't get this, painful, feeling and look on his face
91T	And how have you tried to solve this problem?
92V	(text omitted) but then we we can't, it always gets like the same, he feels bad, then I get very frustrated and I start feeling more bad, it's like, never ending
93T	Mm have you ever managed to have a good talk that you both liked during the last half a year?
94V	Yes after a while. I think it goes like first Alfonso starts starts feeling this pain, then I start feeling bad because of that and after some time, after some crying and some, I don't know for how long that lasts then then we can talk but it's never easy like it's not like (,) I think usually it shouldn't be this hard like you should be able to talk without all this, extra
95T	Mm mm, and it always happened that in the end you managed to have a good talk with each other?

In response to the therapist's initial question (turn 7), Victoria took the lead and marked as the start of the problem a point in time 2 years earlier, when she had become depressed, which 'left scars' in their relationship. After this introduction, Victoria and Alfonso jointly constructed the following account: because of Victoria's past depression, Alfonso now becomes easily impatient whenever Victoria expresses negative feelings; Victoria feels sad and angry that they cannot talk and seeks Alfonso's reassurance; Alfonso reacts with irritation and 'panic'; when Victoria sees this reaction, she feels worse; this cycle 'always' leads to intense arguments that often last for days. This jointly constructed account represents the couple's difficulties in terms of a repeated and problematic interactional cycle, a construction, which is arguably in line with a relational perspective characteristic of couple therapy (e.g. Davis & Piercy, 2007; Sprenkle et al., 2009).

Although the problem was jointly constructed by the couple, they seemed to draw upon different discourses about ideal romantic relationships in their narratives. Drawing upon Taylor's theory of identity (1989), where it is argued that implicit notions about the 'good life' underpin our everyday social action and serve as the benchmark against which we compare our current life and relationships, it has been suggested that moral values often constitute a core issue in therapy talk, although these are not always explicitly recognised as such (McLeod, & Lynch, 2000). From this perspective, Victoria's narrative can be described as centring on the values of connectedness and intimacy and she constructed the problem accordingly, as resulting from lack of closeness, communication problems and Alfonso's divided loyalty and commitment. Her talk reflected a 'romantic-relational' story, a culturally available narrative that values close, intimate relationships with others as the meaning of a good life (Gergen, 1991). Alfonso's description of an ideal relationship, on the other hand, centred on autonomy and independence. As has been argued by narrative theorists (e.g. Angus & McLeod, 2004), these conflicting orientations are common— and arguably gendered—contemporary core narratives.

The problematic interactional pattern constructed by the couple was discussed through the sessions both in general terms and in relation to specific issues, the most important of which concerned the conflicts that arose around Alfonso's visits to his home country. Moreover, in their accounts, the clients underlined a change from 'then' (when Victoria was depressed and Alfonso was understanding and supportive) and 'now' (when Victoria feels better but Alfonso is less patient) and stated that they do not understand why this change has occurred and that they feel unable to deal with it.

It is worth noting that both partners assumed a position of *reduced agency* with regards to their behaviours that had been marked as part of the problematic cycle. Both represented these actions as overly intense and unwanted but as 'automatic' and outside their control (e.g. turns 8, 18, 51, 88, 92). Through the sessions, in the various elaborations of the problem construction, several factors were introduced that served as explanations of and justifications for these 'problem' behaviours, including childhood experiences, subconscious forces, Alfonso's split loyalty to two families, and cultural differences in emotional expression. One way of understanding this persistent nonagentic positioning is in relation to *blame*, which is a common issue in family and couple therapy, often implicated in the construction of the problem (Kurri & Wahlström, 2005; Stancombe & White, 1997, 2005). In these sessions, there were several points of tension, arguments and blaming sequences between the couple; in this interactional context, assuming reduced agency for behaviours they 'cannot help' can be seen as one way the speakers used to refute blame. Blame was more explicitly assigned by Victoria to Alfonso, often through ascribing contradictory positioning, i.e. his behaviour was represented as inconsistent with the position of a 'good', dedicated partner. In addition, although the couple agreed on the problem description, they punctuated their accounts differently. Victoria usually started her description of the problem with reference to Alfonso's 'pained expression', thus marking this as the start of the problematic cycle and the trigger for her response (e.g. turn 90). Accordingly, Alfonso underlined Victoria's oversensitivity to the way he talks, marking her reaction as the start of the problem cycle and his angry outbursts as a response. These differing punctuations constitute subtle blaming sequences, where each partner locates the beginning of the problematic pattern in the other's behaviour, frames their own action as an understandable and justifiable response to this trigger, and locates the solution in the other partner changing.

Shifts in Meaning Construction and Positioning

Through the first three sessions, the pattern described in the initial problem construction was elaborated upon and expanded to include Alfonso's family of origin, the conflicts associated with his relationship to them and particularly to his mother, and cultural differences in emotional expression. Arguably, these discussions opened up space for the clients to express their feelings about each other, to listen to each other's perspective and to explore their expectations from their relationship. Gradually, the narrative about the problem and their relationship became enriched, more complex and encompassing important relational and contextual factors. In this expanded problem formulation, Alfonso's contradictory positioning was 'remedied', as his actions were rendered meaningful in relation to his position as the son of a controlling yet unloving mother.

In addition to the elaboration and expansion of the initial problem formulation, the therapists also put forth a reformulation of the problem at the end of the second session (extract 2). The difficulties the couple faced were reconstructed as part of an expected developmental trajectory involved in setting up a new family, which all young couples face, and complicated because of cultural differences. In this new narrative, each partner's behaviour was represented as an attempt to find a balance between the 'three families' in their life and their conflicts were reframed as evidence of the couple's increased commitment to each other.

In sum, through the sessions the following changes were observed in language use: the problem constructions became more inclusive, complex and emotionally rich, as well as less rigidly held. Broadening the explanatory frame helped in locating the problem actions in a wider context (families of origin and culture) with an effect of diffusing blame. As a consequence, both clients began to assume more agentic positions with regards to their actions and joint life. These changes in discourse use are arguably in line with the literature on common factors, which describes as key aspects of family and couple therapy, the processes of conceptualising client difficulties in relational—rather than individual—terms, disrupting dysfunctional relational patterns and expanding the direct treatment system and the therapeutic alliance (Sprenkle et al., 2009).

Next, we turn to the question of how these shifts were co-constructed by examining in more detail the therapist's talk and formulating the role his utterances may have played in promoting change.

The Therapists' Contributions and Agenda

For this part of the analysis, all the primary therapist's[1] utterances were examined and coded in relation to their form and their function. Interest in the therapist's contributions reflects the assumption that therapists are always active in shaping the course of the conversation, although *how* this is achieved as well as *the effects* the therapist's talk may have may vary. Moreover, power is considered as a constructive force, which together with resistance is always implicated in therapeutic talk, irrespective of the participants' intentions (e.g. Guilfoyle, 2003, 2006). Nevertheless, focussing on the therapist's talk does not imply that his talk has a unilateral or linear

---

[1] The analysis focuses on the primary therapist, given that the co-therapist's contributions were minimal and limited to the reflecting conversations that took place towards the end of the sessions.

effect on the meanings which develop through the sessions. As will be shown in the analysis, clients often resist proposed reconstructions and position calls, and the conversation that unfolds is a joint creation. I am arguing, however, that punctuating on the therapist's talk and examining his utterances in their interactional context can provide a useful point of focus for research.

In this study, the therapist's contributions were studied from a functional and rhetorical perspective using tools from discourse analysis. As already mentioned, the function of a speaker's talk can be deduced through paying attention to its linguistic features and to sequence. In brief, the analysis showed that the therapist was active in shaping the unfolding conversation, although this activity was skilfully subtle. Moreover, most of his utterances seemed to be in line with dialogical principles, as he was minimally intervening, generally responsive to the clients' words, and tended to focus on the clients' personal experiences and meanings. The therapist's discursive agenda could be summed up as serving two main functions: *eliciting narrative elaboration* and *expanding on nondominant narratives*. These are described in more detail in the following sections.

Eliciting Narrative Elaboration

As already mentioned, a shift was observed in the stories told by Victoria and Alfonso about the difficulties they faced, their life together, their hopes and fears for the future and their expectations from each other; their narratives gradually became more complex and rich, emotionally varied and polyphonic. Through examining all instances in the session where the clients introduced new aspects to their narratives or elaborated upon previously introduced themes, it became apparent that narrative elaboration generally followed therapist's utterances, which were minimally intervening. This is in contrast to other studies that show that therapists often elicit narrative elaboration through direct questions, information eliciting tellings or reformulations (Buttny, 2004; Davis, 1986). The main way in which the therapist in this case elicited further narration was through a particular listening stance that contributed to *establish and maintain intersubjective understanding*, as described below. The therapist was minimally intervening, in the sense that the majority of his utterances neither added to the meanings expressed by the clients, nor reframed them. Close analysis shows that he was very active, however, in promoting the process of dialogue and worked hard to create a sense of respectful curiosity about and responsiveness to the clients' words, experiences and meanings. The main ways in which this was effected were through minimal responses, third person repairs and simple evaluations, which were by far the most frequent forms of utterance he used. *Minimal responses* refer to brief paralinguistic responses ('uhmuhm', 'yes', etc.) that aim to communicate the listener's attention and understanding; *third person repairs* refer to brief utterances by the listener that do not add anything new in terms of meaning but display understanding and give the speaker the opportunity to correct a possible misunderstanding or consolidate their point. The therapist used many third person repairs, most often by repeating the clients' words. *Reflections or minimal evaluations* are sequences in which the speaker (client) recounts a story and then the listener (therapist) evaluates it, by selectively underscoring some aspect of the story (e.g. turn 57, extract 1). The therapist most often reflected the clients' talk by repeating word-for-word part of the previous utterance, with a selectively preference for words that describe personal experience and emotions.

In addition to the types of utterance described above, the therapist also contributed to narrative elaboration by *direct questions*, usually aimed at clarifying a point or obtaining further information (e.g. turns 51, 53); he also often used direct questions to invite clients to recount a specific event or a personal experience in detail, thus promoting narratives richer in detail and emotional intensity.

It is worth noting that the therapist seemed to use slightly different responses depending on who he was addressing; he used more minimal responses in his exchanges with Victoria, who was generally more talkative and arguably better conversed in therapy talk, whereas he used more direct questions when addressing Alfonso. This can be seen as a further example of the therapist's responsiveness, in the sense that he seemed to adapt to the clients' different interactional styles, which arguably functioned to promote the therapeutic alliance with both partners.

Despite being subtle, the therapist's utterances arguably had powerful effects in the interaction, as they allowed for emotions to be expressed and for new meanings to gradually develop; in this sense creating a sense of intersubjective understanding, mainly through listening, seems to have promoted narrative elaboration in this case.

Expanding on Nondominant Narratives

The second main aspect of the therapist's agenda involved *exploring nondominant themes and positions* and positions and expanding the interpretative frame, and in this way contributing to shifts in meaning. The therapist contributed to such shifts in several ways described below. A general feature of his talk, however, was that these utterances were offered tentatively and he often invited the clients' comments on them, thus positioning himself on equal footing in the conversation and marking his formulations as open to revision.

The therapist in several instances affected the course of the conversation by highlighting similarities and providing links between differing (and even opposing) accounts. This was achieved, for example by *illuminating interactional patterns,* which moved towards framing behaviours in relational terms (turn 19). Moreover, he sometimes used *expansion questions*, i.e. questions that broaden the frame of understanding problem behaviours, and *reversals*, i.e. statements that subvert the dominant narrative (Kogan & Gale, 1997). Furthermore, he used *questions oriented towards resourcefulness*. For example, in turns 85–95, the therapist's questions attempt to shift the dominant narrative—which represented the couple's 'problem behaviours' as entrenched and themselves as helpless to effect any change—by focussing on attempted solutions (turns 91, 95), preferred outcomes (turn 89) and unique outcomes (turn 93). In addition, the dominant narrative was subverted through questions that positioned the clients as reflexive and agentic (turns 87, 89). It is worth noting, however, that at this early point in therapy these 'position calls' were not always accepted by the clients (e.g. turns 92, 94).

In addition to questions, the therapist sometimes shaped the course of the conversation by *shifting the focus of the conversation*, for example by uttering a minimal agreement and then moving on to a different account or another speaker (e.g. turn 57). The therapist also consistently seemed to dis-prefer and *sidestep* specific types of utterances, mainly problem talk (turn 95).

Less often, the therapist used *direct challenges* and *reformulations*. Challenges were usually delivered in a light, playful and humorous tone. Humour tends to be used by therapists when managing a delicate reformulation or reflection (Buttny, 2004), for example when they formulate something which could be perceived as blaming. For example, Alfonso's talk in turn 58 is characterised by many false starts and vague words, which can be seen as marking this as a delicate issue; the therapist's humorous response (turn 59) seems to help talk about his lack of patience without ascribing blame.

In a few instances, the therapist explicitly introduced a new narrative in the form of reformulation. *Reformulations* are considered powerful discursive tools that are particularly common in institutional conversations and function to advance an institutionally relevant version of the speakers' words. Reformulations selectively focus on one aspect on what has been said by the previous speaker or put a particular spin on it, often changing subtly what has been said, while seemingly accepting it. Therapists routinely use reformulations to recast the clients' troubles in the language of therapy and have been shown to use several different rhetorical strategies to that effect (Antaki, Barnes, & Leudar, 2007; Buttny, 2004; Davis, 1986). In this therapy, reformulations were used rarely; they were built gradually through the session and were designed as limited and open to revision. Perhaps, the clearest example of a reformulation was the reframing of the couple's difficulties as part of a normal developmental trajectory of establishing a new family, which was put forth in the second session, as shown in the following illustrative examples:

Extract 2: Building a reformulation

Session 2, turns 326, 361, 422–430

326T	You've been living together for three years now. Has it changed, your visits to [home country]? And the place of the visits [in your life]?
361T	and now you are three families, you are one family [pointing to Victoria] you have your own family, and you [pointing to Alfonso] have your own family, and, and, I would suppose that it is a bit different compared how it is the first time, how is it and how is it when the situation has become more stable, and longstanding (.) and in a way what has happened is that you have formed a family during these three years, the two of you
422T	It would be complicated even with two Scandinavian families
423V	yeah
424T	being in between and you are
425A	But I think Scandinavian families they are much more easy going

426V   Yeah I think so too
427A   I don't have any problems with her parents, at all, so
428T   Ok, I have seen many difficult Scandinavian families as well
429V   yeah
430T   Well I am a family therapist, so I see a lot, but, but, oh, I don't know easier but at least challenging situations, so I can easily imagine that even in, if you are coming from the same culture, from the same background, you can have quite a challenging situation

An interesting exception to the therapist's usual style regarding authority and expertise occurs in the above extract. The therapist (turns 422, 424) puts forth the idea that it is 'complicated' for any new couple to be 'in between' (thus problematizing this aspect of their joint life) a formulation which the clients jointly resist (turns 425–429). In response, the therapist calls upon his expertise as a family therapist, using a common rhetorical strategy which strengthens the persuasive power of his argument.

The reformulation was then fully articulated in turns 503–505 during the reflecting conversation, as shown below.

503T1   (…) today it has become much clearer that when Alfonso and Victoria are having their own family of two persons now, they are in between two other families, and, and that is challenging for every new and young families to be living, but there is an extra challenge with them, that they are living together with language that is not their own language, Alfonso and Victoria and these two families are living in a language that Victoria cannot understand, and Alfonso said that has learned to speak [Scandinavian language] and perhaps he can follow what people in Victoria's family are speaking about, so that's the big challenge, to do (.) you did not you understand what I was saying..
504T2   I understand
505T1   OK, I didn't understand myself what I was saying [laughter] OK, the thing that I was puzzling, to say the same thing in another way, is that, that in addition (.) to the love that they feel about each other, and of course that is the basic idea in a love relation, I have to feel, as Victoria said, that I am the first person in your life, there is no other, there is no sense to be together if you don't feel it (.) but at the same time there are so many other practical issues, that we have to take into account more, and in my mind it seems to be that this family is having these two other families is a kind, is a, of issue as well

Some interesting aspects of this reformulation, which has been gradually built through the session, are noted briefly. Firstly, when the therapist puts forth the view that the problems the couple are facing are part of the normal course of 'building a new family', his talk is characterised by certainty; there are no markers of uncertainty or tentativeness and his turn is constructed as an objective account. For example, there are no discursive markers that this is a personal account and the 'evidence' is outlined for a situation that has now 'become much clearer'. However, at the end of his turn, the therapist downplays his authority through humour ('you did not understand what I was saying' and 'OK I didn't understand myself what I was saying') thus marking his reformulation as tentative. Then (turn 505), he states that he was 'puzzling' (i.e. curious yet uncertain) and proposes an alternative formulation which he, however, frames as if it is 'the same thing in other words'. Here he first puts forth the clients' perspectives (Victoria's focus on the importance of being the first one in a love relationship and Alfonso's focus on the practicalities that can get in the way) which are stated with certainty and strong language, before restating his own formulation, which is now marked as a personal view and is also worded with less certainty (e.g. in my mind, it seems, is a kind of an issue). In this way, the three differing perspectives are marked as equally valid and as complementary (thus a both/and perspective is created).

In sum, the therapist's agenda could be described as promoting emotional expression and narrative elaboration, diffusing blame and broadening the interpretative frame. These shifts were shown to be achieved primarily through minimal interventions, as the majority of the therapist's utterances were oriented towards establishing intersubjective understanding rather than directly shifting meaning. Moreover, the therapist tended to start the sessions with minimal responses and reflections that oriented towards the expression of feelings and the detailed description of personal experience. In the middle phase, he increasingly used evaluations and questions inviting reflection and further meaning elaboration and it was not until the latter part of the session that he used (albeit rarely) more strongly intervening utterances, such as reformulations.

Discussion and Conclusions

The process of change in this couple therapy can be broadly seen to involve shifts in meaning construction, in positioning and in interaction. More specifically, the construction of new relational meanings, the reduction of blamings and the negotiation of more agentic subject positions were shown to be implicated in the process of change. These findings are in line with the research literature on the therapeutic factors in couple therapy (Sprenkle et al, 2009).

Moreover, the therapist in this case, although generally nonintervening, was very active in creating a space where a particular kind of conversation could take place. A primary effect his talk had on the unfolding conversation was the expression of feelings, the elaboration of the clients' narrative, the promotion of relational accounts, and increasing focus on resourcefulness and agency. His interactional style could be described as primarily oriented towards creating space for dialogue, in line with dialogical and collaborative approaches to family therapy (Anderson, 2012; Seikkula, 2011). As such, this analysis illuminated some of the interactional processes involved in working dialogically with a couple.

With regards to the usefulness of the method for process research, discourse analysis can illustrate the linguistic and discursive phenomena that underlie psychotherapeutic processes and thus provide bridges between the macro-level of therapy theory and the micro-level of clinical interaction. Moreover, focussing on the therapist's interventions can enhance therapist's creativity and reflexivity; attending to such issues raises issues of authorship, power and authority and makes the adoption by the therapist of an informed and ethical position even more pressing (Avdi &

Georgaca, 2007a). Several criteria for the quality of discourse analysis have been proposed, including internal coherence, rigour, transparency and situatedness of the analysis, reflexivity and usefulness (Georgaca & Avdi, 2012). In this chapter, I have tried to support the analytic claims through the discussion of illustrative extracts and to make the analysis transparent, although no claims are made about the 'truth' of this particular analysis; indeed, the aim of discourse analytic research is not to provide a definitive reading of the material but to promote further dialogue.

In sum, in this chapter I have argued for the usefulness of discursive research for providing family therapists with useful concepts and tools to study talk-in-interaction. Discursive analyses can help examine therapy as a joint dialogical achievement, furthering our understanding of the relational and intersubjective processes through which family therapy gets done. Moreover, many of the skills required for conducting such research are not unlike clinical skills, and so small-scale, detailed idiographic research on therapy sessions could form part of therapists' work, helping towards bridging the gap between research and clinical practice, and fostering therapist reflexivity.

References

Anderson, H. (2012). Collaborative relationships and dialogic conversations: Ideas for a relationally responsive practice. *Family Process, 51*(1), 8–24.

Anderson, H., & Goolishian, H. A. (1988). Human systems as linguistic systems: Preliminary and evolving ideas about the implications for clinical theory. *Family Process, 27*, 371–394.

Angus, L. E., & McLeod, J. (Eds.). (2004). *The handbook of narrative and psychotherapy: Practice, theory and research*. London: Sage.

Antaki, C., Barnes, R., & Leudar, I. (2007). Members' and analysts' analytic categories: Researching psychotherapy. In A. Hepburn & S. Wiggins (Eds.), *Discursive research in practice: New approaches to psychology and interaction* (pp. 166–181). Cambridge: CUP.

Avdi, E., & Georgaca, E. (2007a). Discourse analysis and psychotherapy: A critical review. *European Journal of Psychotherapy and Counselling, 9*(2), 157–176.

Avdi, E., & Georgaca, E. (2007b). Narrative research in psychotherapy: A critical review. *Psychology and Psychotherapy: Theory, Research and Practice, 80*, 407–419.

Barkham, M., Stiles, W. B., Lambert, M. J., & Mellor-Clark, J. (2010). Building a rigorous and relevant knowledge-based for the psychological therapies. In M. Barkham, G. Hardy, & J. Mellor-Clark (Eds.), *Developing and delivering practice-based evidence: A Guide for the Psychological Therapies* (pp. 21–61). London: Wiley.

Bertrando, P. (2007). *The dialogical therapist*. London: Karnac.

Buttny, R. (2004). *Talking problems: Studies on discursive construction*. Albany, NY: State University of New York Press.

Davies, B., & Harré, R. (1990). Positioning: The discursive production of selves. *Journal for the Theory of Social Behaviour, 20*(1), 43–63.

Davis, K. (1986). The process of problem re(formulation) in psychotherapy. *Sociology of Health and Illness, 8*, 44–74.

Davis, S. D., & Piercy, F. P. (2007). What couples of therapy model developers and their former students say about change, part II: Model-independent common factors and an integrative framework. *Journal of Marital and Family Therapy, 33*, 344–363.

Elliott, R. (2012). Qualitative methods for studying psychotherapy change process. In D. J. Harper & A. Thompson (Eds.), *Qualitative research methods in mental health and psychotherapy: An introduction for students and practitioners* (pp. 69–82). Chichester, UK: Wiley.

Georgaca, E., & Avdi, E. (2012). Discourse analysis. In D. J. Harper & A. Thompson (Eds.), *Qualitative research methods in mental health and psychotherapy: An introduction for students and practitioners* (pp. 147–162). Chichester, UK: Wiley.

Gergen, K. (1991). *The saturated self: Dilemmas of identity in contemporary life*. New York: Basic Books.

Guilfoyle, M. (2001). Problematizing psychotherapy: The discursive production of a bulimic. *Culture & Psychology, 7*, 151–179.

Guilfoyle, M. (2003). Dialogue and power: A critical analysis of power in dialogical therapy. *Family Process, 42*(3), 331–343.

Guilfoyle, M. (2006). Using power to question the dialogical self and its therapeutic application. *Counselling Psychology Quarterly, 19*(1), 89–104.

Hak, T., & de Boer, F. (1995). Professional interpretation of patient's talk in the initial interview. In J. Siegfried (Ed.), *Therapeutic and everyday discourse as behavior change: Towards a micro-analysis in psychotherapy process research*. Norwood: Ablex.

Harré, R., & Van Lagenhove, L. (Eds.). (1998). *Positioning theory*. Oxford: Blackwell.

Henriques, J., Hollway, W., Urwin, C., Venn, C., & Walkerdine, V. (1998). *Changing the subject* (2nd ed.). London: Routledge.

Hermans, H. J. M., & Dimaggio, G. (2004). *The dialogical self in psychotherapy*. London: Brunner-Routledge.

Kogan, S. M., & Gale, J. E. (1997). Decentering therapy: Textual analysis of a narrative therapy session. *Family Process, 36*, 101–126.

Kurri, K., & Wahlström, J. (2005). Placement of responsibility and moral reasoning in couple therapy. *Journal of Family Therapy, 27*, 352–369.

Lambert, M. J., & Ogles, B. M. (2004). The efficacy and effectiveness of psychotherapy. In M. J. Lambert (Ed.), *Bergin and Garfield's handbook of psychotherapy and behavior change* (5th ed., pp. 139–193). New York: Wiley.

McLeod, J. (2011). *Qualitative research in counselling and psychotherapy* (2nd ed.). London: Sage.

McLeod, J., & Lynch, G. (2000). 'This is our life': Strong evaluation in psychotherapy narrative. *European Journal of Psychotherapy, Counselling and Health, 3*(3), 389–406.

Norcross, J. C. (2011). *Psychotherapy relationships that work: Evidence-based responsiveness* (2nd ed.). Oxford: Oxford University Press.

Salvatore, S., & Gennaro, A. (2012). The inherent dialogicality of the clinical exchange: Introduction to the special issue. *International Journal of Dialogical Science, 6*(1), 1–14.

Seikkula, J. (2011). Becoming dialogical: Psychotherapy or a way of life? *The Australian and New Zealand Journal of Family Therapy, 32*(3), 179–193.

Seikkula, J., & Arnkil, T. E. (2006). *Dialogical meetings in social networks*. London: Karnac.

Seikkula, J., & Olson, M. E. (2003). The open dialogue approach to acute psychosis: Its poetics and micropolitics. *Family Process, 42*(3), 403–418.

Seikkula, J., & Trimble, D. (2005). Healing elements of therapeutic conversation: Dialogue as an embodiment of love. *Family Process, 44*(4), 461–475.

Sprenkle, D. H., Davis, S. D., & Lebow, J. L. (2009). *Common factors in couple and family therapy*. London: Guilford Press.

Stancombe, J., & White, S. (1997). Notes on the tenacity of therapeutic presuppositions in process research: Examining the artfulness of blamings in family therapy. *Journal of Family Therapy, 19*, 21–41.

Stancombe, J., & White, S. (2005). Cause and responsibility: Towards an interactional understanding of blaming and 'neutrality' in family therapy. *Journal of Family Therapy, 27*(4), 330–351.

Stiles, W. B., Honos-Webb, L., & Surko, M. (1998). Responsiveness in psychotherapy. *Clinical Psychology: Science and Practice, 5*(4), 439–458.

Stiles, W. B., & Shapiro, D. A. (1994). Disabuse of the drug metaphor: Psychotherapy process-outcome correlations. *Journal of Consulting and Clinical Psychology, 62*, 942–948.

Strong, T., Busch, R. S., & Couture, S. (2008). Conversational evidence in therapeutic dialogue. *Journal of Marital and Family Therapy, 34*, 288–305.

Taylor, S. R. (1989). *Sources of the self: The making of the modern identity.* Cambridge, MA: Harvard University Press.

Wampold, B. E. (2001). *The great psychotherapy debate: Models, methods, and findings.* Mahwah, NJ: Lawrence Erlbaum.

Wetherell, M., & Edley, N. (1999). Negotiating hegemonic masculinity: Imaginary positions and psycho-discursive practices. *Feminism & Psychology, 9*, 335–356.

Wetherell, M., Taylor, S., & Yates, S. J. (Eds.). (2001). *Discourse theory and practice: A reader.* London: Sage.

White, M., & Epston, M. (1990). *Narrative means to therapeutic ends.* New York: Norton.

Chapter 7
Dominant Story, Power, and Positioning

Helena Päivinen and Juha Holma

From a narrative perspective, how we understand our life, experience, and problems is shaped and defined by cultural discourses (Dickerson & Crocket, 2010; White & Epston, 1990; Winslade, 2005). In these cultural discourses, power is also present, as these discourses constitute our "truths" and normality (Foucault, 1977). Love relationships are also embedded in these discourses. In psychotherapy, what constitutes normality and truth in love relationships is constructed via discursive practices. These psychotherapy discourses in turn are affected by larger cultural discourses.

Narrative psychotherapy research has been conducted from a wide variety of approaches and has focused mainly on client micro-narratives (Avdi & Georgaca, 2007b). Along with Avdi and Georgaca (2007b), we see a need for narrative studies that acknowledge the broader sociocultural processes involved in narrative production and transformation. In this study, we look at narratives from one such broader, discursive perspective. The analytic tools that we have chosen allow us to place the stories that are constructed in psychotherapy within these larger cultural discourses. We investigate how dominant stories are constructed in couple therapy sessions in the interaction between the participants, including the therapists. We study how participants in couple therapy construct and reconstruct their positions and focus on the resulting distribution of power or the rights and duties that these positions entail. We are also interested in the way people make changes in identity by differently positioning themselves and others (Winslade, 2005).

H. Päivinen (✉) • J. Holma
Department of Psychology, University of Jyväskylä, Jyväskylä, Finland
e-mail: helena.paivinen@jyu.fi

© Springer International Publishing Switzerland 2016
M. Borcsa, P. Rober (eds.), *Research Perspectives in Couple Therapy*,
DOI 10.1007/978-3-319-23306-2_7

Review of the Literature

Narrative Approach and Dominant Story

The narrative approach postulates that a narrative or story offers a frame in which lived experience is structured. It is in stories that we situate our experience and give meaning to our experience. A narrative or a story has a beginning, middle, and an ending as well as a plot that holds the story together (Sarbin, 1986). Stories enable persons to join aspects of their experience through the dimension of time (White & Epston, 1990). In this way, the world and they themselves become coherent, which invests life with a sense of continuity and meaning. Thus, stories are constitutive and shape our lives and relationships.

The stories people tell about their lives reflect the prevailing cultural narratives (Bruner, 1987, 1991). During these tellings, some stories become dominant while others become silenced (McLeod, 2004). These storied dominant narratives reflect the culturally dominant discourses and values in the teller's society (Hare-Mustin, 1994; McLeod & Lynch, 2000) and therefore often seem to be appropriate and relevant when we want to express our experience (White & Epston, 1990).

Nevertheless, a story, even a dominant one, can only partly cover the individual's lived experience (Bruner, 1986). When the dominant story does not sufficiently represent the individual's lived experience or when there is contradiction between the dominant story and the lived experience, the person may experience unease, and seek therapy (White & Epston, 1990). Getting stuck with problematic stories has been seen to affect relationship problems (Rosen & Lang, 2005; Sween, 2003; Zimmerman & Dickerson, 1993). In their paper, Sinclair and Monk (2004) illustrate how couple conflicts can be approached discursively in couple therapy; this, according to them, leads to a more open and less blame-attributing therapy practice.

Power in Psychotherapy

Cultural discourses are present in forming the stories narrated about and the positions taken in love relationships. Feminist family therapists argue that owing to patriarchal discourse there is always an imbalance of power in heterosexual relationships (Dickerson, 2013). However, at least in our Western culture equity between the partners is usually a premise in love relationships and in couple therapy. Others have approached inequality in heterosexual couples by concentrating on mutuality for which, they argue, cultural discourses are not properly established (Knudson-Martin, 2013; Knudson-Martin & Huenergardt, 2010).

According to Foucault (1977), power works in psychotherapy and other psychological domains through taking people's minds as objects of professional knowledge. It is argued that therapists are participants of a specific kind, as their institutional position offers them more influence and significance than the family members

(Guilfoyle, 2001). The cultural and psychological discourses that are used in meaning making in psychotherapy are limited to those that are familiar to client and therapist (Hare-Mustin, 1994; Parker, 1998). If therapists are not aware of their position and the possible oppressive nature of the cultural discourses that are drawn on, there is a risk that social injustice and oppression will be reproduced in the therapy room (Dickerson, 2013; Hare-Mustin, 1994).

Discourse Analysis and Positioning

There are significant differences within the discourse analysis tradition (Avdi & Georgaca, 2007a; Parker, 2013; Potter, 2004). However, what these approaches share is their interest in the use of language in discourse. Discourse is seen as active and purposive in constructing realities. Parker (2013) states that discourse analysis should not be conducted by strictly following certain methodological steps. Instead, new ways of combining analytical tools should be devised. In this present study, we use the concepts of position and positioning as our main discourse analytic tools. Positioning can be described as the discursive process in which people are given parts or are assigned locations in the discourse (Davies & Harré, 1990; Hollway, 1984). Position is a central concept in studying discourse (Sinclair, 2007). It has particular value in studying the detail of how discourse operates in the production of relationships (Winslade, 2005) and in the changing responsibilities and interactive involvements of the members of a community (Linehan & McCarthy, 2000).

Positioning occurs within the context of a specific moral order of speaking (Harré & van Langenhove, 1991; Chap. 10) and, as well as reflecting cultural discourses, these local stories also reflect the positions embedded in these discourses. The concept "position" focuses on the dynamic aspects of encounters (Davies & Harré, 1990). Positions change and shift as we draw from the various cultural discourses available. Positioning can also be resisted, which also changes the discourse. Furthermore, a change of positioning, and hence in the discourse, can be accomplished by repositioning: taking up or offering another new position.

Positions are always relational (Harré & Moghaddam, 2003; Hollway, 1984; Winslade, 2005). This means that the act of positioning someone in a discourse inevitably entails positioning the other participants relative to that initial positioning. In our analysis, we use the term "counter position" to refer to these entailed positionings to highlight the relational nature of positioning. Positioning may be explicit or implicit. Explicit positioning refers to deliberately constructing certain uttered positions while implicit positioning occurs when positions are taken for granted and thus they are not uttered as explicitly. In either case, positions set the limits of what are considered socially and logically possible actions in the discourse (Harré & Moghaddam, 2003; Hollway, 1984). In other words, one's position defines one's rights, duties, and obligations and thus the distribution of power between the participants in a discourse.

Data Analysis and Method

In this present study, we analyze the course of four therapy sessions attended by Victoria and Alfonso, a young, multicultural couple. We focus on how dominant stories are constructed in these sessions and how power is distributed in the positions entailed by these dominant stories. Our analysis started with multiple readings through the transcribed therapy sessions. Next, a narrative approach was used to organize the experience of the clients into stories. Our units of analysis were the dominant stories constructed in the therapy discourse. A dominant story was identified as a narrative that was accepted by all the other participants in the therapy session. That means none of the others participating in the conversation criticized that particular narrative. The dominant stories were identified and marked in the transcripts. The evolution of these dominant stories and new emerging stories was followed throughout the therapy process. The stories identified as dominant were studied using positioning as a discursive tool of analysis. We looked at what kinds of positions were constructed for the clients and what kinds of power aspects were entailed by these positions.

Findings

We illustrate the construction of each dominant story as it evolved in the couple therapy sessions. The findings are presented with selected data extracts in which the story construction and the positioning in each dominant story become explicit.

Session 1

Dominant Story 1

At the outset of the therapy session, Victoria relates how in the past she was depressed and how this greatly affected their relationship. At that time, Alfonso was a caregiver, supportive, and understanding towards Victoria. Now, the couple see that the situation has changed. Victoria says she is feeling better but that she still has a strong need to talk. Alfonso says he does not have the patience to listen like he used to when Victoria was depressed.

Extract 1

Session 1, turns 18–22

18 A   But then I think that I, somehow I react too strongly like if it would be something bigger

19 T   So you are saying that meanwhile Victoria was down you could support her but now when things are better, in some way,}

20 A   Yeah, now if there is something small maybe that somehow I feel I kind of get like how'd you say aaa irr}
21 A   } Irritated? Yeah?
22 V   } Like he's afraid that every time I need to talk about something, something a bit negative or whatever that it will get again to this kind of awful situation where I am crying in bed for 2 days like that, kind of, he doesn't believe kind of that that I'm better and that I I'm able to talk about things and that I really need to talk about things.

Victoria positions herself as free talker and by this act she constructs for Alfonso the counter position of being a patient listener. Alfonso is not satisfied with this position but becomes irritated when Victoria wishes to talk about something. Victoria relabels Alfonso's emotion fear and emphasizes that there is no reason to be afraid, since she is strong enough to talk about things without collapsing. In this way Victoria constructs her behavior as unproblematic. The relabeled emotion fear is later accepted by Alfonso himself and by the therapist. The current problem is constructed as Alfonso being afraid and not able to talk but instead reacting with panic and anxiety when Victoria starts a conversation on certain topics. This leads the couple to quarrel.

Victoria and Alfonso validate their positions by reference to their needs. Alfonso needs to rest and Victoria needs to talk.

Extract 2

Session 1, turns 74–80

74 A   } I think it can be because like, I, somehow now maybe now that she got better, so somehow I have used a lot of, let's say energy, maybe I kind of feel now that, [Laughter] I kind of need a rest
75 T   You need a rest, yeah
76 A   I think somehow, I mean that's the explanation that I
77 T   You have used so much energy that now you need rest
78 A   Yeah, I think, like I would need … some positive experience, like that if for some time there wouldn't be this kind of discussion, then maybe we could then (.) get like normal
79 T   Yeah
80 V   but this is difficult because in a relationship there has to be talking about things if there is anything that bothers me I need to share it, I need to talk about it but since that's impossible

Extract 3

Session 1, turns 91–93

91 T   And how have you tried to solve this problem?
92 V   We have been talking about it like we both know the situation like, we know why is it like this and we think that time would make it better but the problem seems to be that, as long as we can't talk how can anything get better? There is always something in a relationship that you need to talk about, but then we we can't, it always gets like the same, he feels bad, then I get very frustrated and I start feeling more bad, it's like, never ending
93 T   Mm have you ever managed to have a good talk that you both liked (V: yes) during the last half a year?

Alfonso positions himself as exhausted, a person in need of rest because his earlier position as caregiver has demanded a lot of energy. By positioning himself in this way, he constructs for Victoria the counter position of a sympathizer, which Victoria has not taken up, as she is not giving Alfonso a positive discursive experience without arguing. Victoria strongly repositions herself as a free talker. This position is no longer explained as a result of depression but as part of a normal relationship. Victoria here refers to a cultural discourse on what constitutes a normal love relationship: "there has to be talking." She also constructs her emotions as something one has to deal with by sharing and talking.

Alfonso's new positioning and the story he refers to are not accepted by Victoria. Instead, the normality that is constructed reflects the cultural discourse that talking is important in love relationships. This discourse prescribes to the couple the positions of a free talker and a patient listener. In Victoria and Alfonso's relationship this means that Victoria talks and Alfonso listens. Thus, Alfonso's inability to talk is constructed as problematic. This problem formulation also fits in with culturally shared discourse of psychotherapy as a talking cure where the participants share and reveal their thoughts and feelings.

The therapist supports Alfonso in constructing his story about the situation. However, later the therapist accepts the formulation of the problem according to the dominant story constructed by Victoria by asking how the couple have tried to solve the problem, as well as the implied cultural discourse by asking whether they have ever managed to have a good talk. Later, when the therapist asks Alfonso's opinion, Alfonso shares this problem formulation and accepts the dominant story. Alfonso would like to be able to listen but says that he does not decide to react in the way he does. The dominant story that in a good love relationship you have to share your experiences and feelings with your partner is now constructed and shared, although it assigns Alfonso the position of a listener, which he is not satisfied with.

Dominant Story 2

In the first therapy session, another dominant story also emerges. Alfonso's visits to his home country create a problem for the couple. Victoria says that she feels that they are not a couple when Alfonso is abroad visiting his family. Victoria has been asking Alfonso to contact her every day when he is away so that she would know that he is thinking of her.

Extract 4

Session 1, turns 133–138

133 V   … I think that the relationship should be, the most important thing in your life or at least one of the most important things
134 T   Yes
135 V   I think it should be the number one because otherwise what's the point

136 T   Mmmm
137 V   And, somehow, I feel like that, still when he is there alone, I feel like, I'm also
          very alone I feel like kind of like we're not together while the time he's there, I
          would need something like one text message like per day sent without my me
          asking for it, but he doesn't need that, he doesn't need to contact me, like
138 A   Well I

Victoria validates her request for messages by pointing out that the love relationship should take first priority in one's life. Here she reflects a cultural discourse about the special value attached to a love relationship. The therapist signals agreement with this cultural discourse. Victoria says the relationship is the most important thing for her, but not for Alfonso. By this utterance she positions herself as a fully committed partner in their relationship and Alfonso in the counter position of an uncommitted partner.

Extract 5

Session 1, turns 140

140 A   Yes for me just that, for the thing that I'm there just a few, few weeks per year and
          in (home country) it's not like this, there's all the family and there's many people
          that you always have to go to see and to meet, to meet your friends or, so when
          I'm there I'm really just busy and anyway at the end I always like manage to send
          a message or to call

Alfonso tries to resist this counter positioning as an uncommitted partner by referring to his other obligations when visiting his family. He says that being in his home country is different and he is busy there. When talking about his visits to his home country, Alfonso refers to a cultural discourse related to the concept of family relationships and their value. In his home country, one's relationships with one's family members are very important. Alfonso repositions himself as a son of his family of origin and also as a committed partner, stating that he does want to keep in contact with Victoria, and that he does in fact do this. The repositioning he proposes is not accepted by other participants in the therapy session; neither by Victoria nor by the therapists. Victoria sticks to her story and the therapists do not actively take part in the story construction at this point. The discussion continues with Victoria positioned as a committed and Alfonso as an uncommitted partner in their relationship.

The division of rights and duties becomes explicit over the session and is illustrated in the following extract.

Extract 6

Session 1, turns 203–206

203 T   and what about if you are the first one sending sms saying 'hello how are you?'
204 V   Yes that's fine yeah but but mostly because we always have to talk about this thing
          I feel I have to kind of see if he does it or not
205 T   Uhuh can you say something more about why, why is this?
206 V   Kind of like testing or something

Victoria says she is testing Alfonso to know whether he is thinking of her. From her position as a committed partner Victoria has the right to ask for proof of this kind. From the counter position of an uncommitted partner, Alfonso has the duty to prove his commitment, which Victoria sees as uncertain.

Session 2

Dominant Story 2

In the second session, there is a return to dominant story 2. Alfonso's visits to his home country are discussed, as Alfonso will be going there soon. Alfonso is planning to send Victoria text messages just as she has asked him to. He asks for Victoria's understanding if for some reason he is unable to send her a message.

Extract 7

Session 2, turns 162–163

162 A no, I told you that, I can send you like I told you, and there are like a week, if there is during this week one day, even if we call, for one day maybe, it can be that or maybe not, it can be that (.) I am really busy (.) I can like do it, I can send you message I can do this, but if for one day it will happens once, like that maybe I can't send you a message (.) once (.) I don't, maybe if you try to understand a little, it can just happen

163 V Yeah, I try (.) I just think that it is quite easy thing to do, if you want to

Extract 8

Session 2, turn 234

234 V But, also I don't want it to feel like work, like I think it should be a bit natural, I don't know, maybe I feel like you, you are not really committed or something because to me it's natural that, I am like interested in what's going on with you or some-

Alfonso attempts to reposition himself as a committed partner despite the fact that he might not be able to send Victoria a text message every day. Alfonso's repositioning is not accepted by Victoria. By uttering that keeping in contact is possible even when busy, if one wants to, she is referring to the cultural discourse of the priority of a love relationship over other family relationships.

Victoria positions herself as committed to their love relationship. She says this commitment comes natural to her. In uttering "natural" she is referring to a cultural discourse that assigns priority value to the love relationship, a discourse that is dominant in her home country, which is also where the therapy meeting is taking place. Finally, Alfonso accepts this cultural discourse and the corresponding duty to send messages. By agreeing to send messages, Alfonso positions himself as a committed partner in the love relationship with Victoria.

Session 3

Dominant Story 2

At the beginning of the third therapy session, there is a brief return to dominant story 2. Alfonso reports that, as agreed, he has sent Victoria text messages during his visit home, and Victoria agrees that keeping contact has worked well. Alfonso has acted in accordance with the cultural discourse brought forward by Victoria, and in this way the problem has been solved. Alfonso is no longer positioned as uncommitted, as he seems to have accepted the discourse of the special value of the love relationship that is dominant in Victoria's home country.

Dominant Story 1 Starts to Evolve

In the session 3, the conversation shifts back to dominant story 1, i.e., in a good love relationship experiences and feelings have to be shared. Victoria says, at the start of this conversation, that Alfonso is continuing to react negatively to the basic premise of dominant story 1. Alfonso's reaction is then explored by the therapists. From their institutional position, the therapists actively ask detailed questions to help the couple in reconstructing the dominant story that has been constructed during the therapy. By doing this exploration, they accept the dominant story constructed by Victoria and acted in accordance with the dominant discourse of the psychotherapy world: experiences and feelings have to be shared.

Extract 9

Session 3, turns 144–148

144 V	like usually it's really some simple question that I would need like one word for an answer, but then I don't get it, I get only this awful, like (.) this very bad reaction
145 T2	What kind of a reaction those are? What do you mean by that?
146 V	Alfonso's reaction is like, his face gets like this and like, I don't know, I think I have explained it but I don't know if you were here(.) but he gets like really suffering (..)
147 A	Yeah, it's a bit like, when you are kind of disappointed, you are a bit down, a bit
148 V	and then for very small reasons I think this happen like, like I think that in every relationship there is times that you, you want to talk about your relationship, it doesn't work like if you never talk about it, and even if I try to talk about positive things (.)

Victoria defines Alfonso's reaction as apparent feelings of suffering. Alfonso himself relabels his emotions as disappointment and feeling down. Disappointment has an object, and hence by saying this Alfonso is assigning the blame to Victoria and resisting the position Victoria is constructing for him. Victoria strengthens her position as a free talker by referring to cultural discourse about the importance of talking about the relationship: "it doesn't work if you never talk about it."

A little later in this session, Alfonso starts reframing his reaction. He says that it is kind of opposite to Victoria's feelings of depression, and at the same time he says that it is the same: like she could not control her depression then, he can not control his reaction now.

Extract 10

Session 3, turns 164–166

164 A   it's, in a way I have this reaction and I (.) can't like help it, … it just comes like this I don't can't control it, and then I can't, don't want to have this kind of reaction but it's just how it is, I don't control it (.) so I was trying to explain to her that this kind of thing, as it would be the exact kind of opposite of how it was for you, that you had this reactions, strong reactions that you couldn't, you couldn't control, and you were just feeling like bad for a long time so

165 T   I don't quite follow you, when you said that first Victoria was feeling bad and you was ok but now it is the opposite you are feeling

166 A   … I think that now maybe when I have this reaction that it's not exactly the opposite because I think that she can't deal with it

Here Alfonso positions himself as a person in need of understanding. Through this positioning, he invites Victoria to take the counter position of a sympathizer. Alfonso validates his position by referring to his uncontrollable emotions.

Extract 11

Session 3, turns 179–181

179 A   Because I was thinking like when you were getting, when you were getting like this kind of things and I was understanding you and I was kind of, there like, so I was just trying to say so if you could try to think it as the opposite way, that what I do it's not something that I decide to do maybe (.) that time if you could understand me like

180 V   but there is the sadness because, because all this happened, because I got like sick, like depressed (.) and now it has ruined everything, that I can never get it back just because I got this, because

181 A   but sadness it's like, of course, I was feeling sad too, it wasn't about, I was feeling sad also when you were having this, this thing, but I think that when you feel like that it's more than just sadness, it's like when you start crying, and then, when I have the thing it gets again to the other side…

Alfonso continues to construct for himself a position as a person in need of understanding, and for Victoria a counter position as a sympathizer. Victoria resists the constructed counter position of a sympathizer by speaking of her feelings of sadness, and now again refers to her depression, first as a justification for her feelings of sadness that prevent her from taking up a sympathizer position, and second as something that has affected their relationship in a negative manner. Alfonso sticks to his view that he too needs understanding. Thus, the couple are negotiating who is allowed to show feelings, to be weak and to be understood.

Next, the therapist frames the issue as expectations of one another and describes how expectations in a relationship change and develop. Victoria accepts this framing but she also sticks to her position as talker and constructs it as problematic that

Alfonso refuses to take up the counter position of a listener. Victoria refers to her individual therapy as validation of her ability to talk and to cultural discourse about talking as central in love relationships as validation of the dominant story.

The therapist takes up an active position and turns the discussion back to Alfonso's reaction, asking whether it has occurred on some other occasion in his life.

Extract 12

Session 3, turns 258–262

> 258 A   Yeah, that's maybe always can be, in some situations of course, in different situa-
>   tions, but in anyways, in situations of, feeling bad or anger or something like that
> 259 T   Yeah.
> 262 A   And now we understand that maybe can be, how, how I have been like, in my fam-
>   ily, in [home country] can be a little bit different than in [Victoria's home country]
>   I am not sure but (.) I was thinking I don't know, when I was young I, sometimes,
>   maybe, aaah, how to say, my mother was hitting us, not like badly but [laughs]

Alfonso connects his reaction to feeling bad or angry. He tells about his background with experience of domestic violence and growing up in another culture. In this way, Alfonso's position as a person with uncontrollable emotions and who needs understanding gets validation. Victoria confirms Alfonso's childhood experience as serious after this extract and this way takes up the counter position as sympathizer that Alfonso has offered her.

It is at this point that the dominant story "in a good love relationship both parties have to share their experiences and feelings" evolves: Alfonso is sharing his feelings and experiences and thus obtains validation of his feelings and behavior. This new dominant story offers Victoria the counter position of a sympathizer.

Later in the session Alfonso's reaction to Victoria starting a conversation is analyzed further. Victoria states that in general she responds to Alfonso's negative reaction by crying. The therapist points out that they have constructed a circle with each responding to the response of the other and emphasizes that Victoria and Alfonso have learned how to let things calm down before discussing something. In this way, the therapist validates the new dominant story and positions.

Session 4

In the fourth session, the couple report that they have not quarreled during the previous weeks and that for both of them life has been good. Alfonso says there has been the usual amount of questioning by Victoria and reacting by him. Nevertheless, fewer situations have led to an argument.

Victoria is no longer accusing Alfonso of being unable to listen to her or of being uncommitted. Instead, Victoria constructs the current problem as her individual project. She explains her reactions by reference to her emotions, which arise from the old way of thinking, not from Alfonso's behavior. Alfonso says he too has tried to take it easier, but that there has not been much need to do this, since Victoria has made things easier. The therapist suggests that the couple have learned to do something different together and both partners accept this suggestion. They seem to be satisfied with their relationship and decide to end their couple therapy.

Discussion

In this study, we looked at the construction of dominant stories in couple therapy and aspects of power in the positions constructed for clients in these dominant stories. Furthermore, the evolution of these dominant stories was studied. Two dominant stories were constructed in the case of Victoria and Alfonso. The first dominant story stated that in a love relationship partners should share the experiences and feelings with one another. Victoria positioned herself as a free talker, and by doing so she positioned Alfonso in the counter position of a patient listener. In the second dominant story, Victoria positioned Alfonso as uncommitted to their relationship, as he did not keep in contact as often as Victoria would have wanted. This second dominant story about the special value of the love relationship counter-positioned Victoria as a committed partner

In the course of the couple therapy, the two dominant stories evolved such that both Victoria and Alfonso were able to accept the positions they had taken or been given in the story. The first dominant story evolved when Alfonso started to behave according to the cultural discourse subscribed to by both Victoria and by the psychotherapy world. He started to share his experiences and feelings and by doing so gained acceptance to his positioning of himself as a person in need of understanding. This also changed Victoria's position from being the only one who needs to be supported and listened to into a more equal partner who in turn has to listen to her partner's experiences and feelings. The way this story evolved increased the level of equality between the partners. The second dominant story evolved when Alfonso accepted the cultural discourse to which Victoria subscribed and which was dominant in her home country, where the couple are living. Alfonso adapted his behavior accordingly to that cultural discourse and by doing so positioned himself as a committed partner in this love relationship.

Cultural discourses are normative about life and relationships (Foucault, 1977). Our results showed how the power of cultural discourses works in couple's dominant stories. The importance of speaking of one's emotional experiences is a well-accepted discourse in psychotherapy and in western culture generally (Parker, 1997). In this couple's therapy, Victoria referred to this cultural discourse, and her story in line with it became dominant in the couple's discussions. Emotion talk has been found to construct, for one person, the privileged position of laying down the rules of the relationship (Kurri & Wahlström, 2003). This person's emotions and needs then become the focus of the relationship and the responsibility for these states is attributed to the other party (Silverstein, Buxbaum Bass, Tuttle, Knudson-Martin, &

Huenergardt, 2006). In Victoria and Alfonso's therapy, Victoria talked about her and Alfonso's emotions from the beginning. From her position as a free talker, Victoria expected Alfonso to be there to listen to her and sooth her emotions. Moreover, naming and relabeling emotions can be used as a means in positioning and repositioning either oneself or another person (Parrott, 2003), and this seemed to work in the construction of the dominant story in Victoria and Alfonso's therapy.

The first dominant story only started to evolve when Alfonso had begun acting in line with it: He started sharing his experiences and emotions. When asked by the therapist, Alfonso told about his childhood experience which led his reaction be seen in a new light. Victoria stressed the seriousness of Alfonso's experience of domestic violence. This view is in line also with northern European cultural discourse as well as with the view of the psychotherapy world. Trauma discourse is strongly embedded in our everyday understanding of how certain experiences affect us (Parker, 1997). In the psychoanalytic psychotherapy discourse, childhood trauma is understood to explain and moderate our present behavior (Parker, 1998). Victim positioning has even been found to be an effective way to avoid responsibility in therapy discourse (Partanen & Wahlström, 2003). Even though Alfonso was mitigating his experience, Victoria seemed to accept this psychological and cultural discourse about trauma and the positions it offered Alfonso and her. These new positions in the story line offered Alfonso and Victoria the possibility to take turns in being understood and sympathizing, talking and listening.

Less polarized and more flexible positions have also been reported at the end of the family therapy by others (Frosh, Burck, Strickland-Clark, & Morgan, 1996). Frosh et al. suggest that this greater equality may lead clients to a better understanding of one another's perspectives and emotional states. In Victoria and Alfonso's therapy discourse, the evolution of the first dominant story and the potential positions of mutual support were in some way already present in the earlier sessions. However, at those stages, these aspects remained more in the background of the therapy discourse, as some stories are always left unheard in psychotherapy (McLeod, 2004). For a dominant story to evolve or to change, validation through negotiation is needed. It has been argued that cultural discourses of equity, such as mutual support and sharing, do not have as strong an effect on heterosexual relationships as do, for example, patriarchal discourses that are characterized by an unequal power arrangement (Knudson-Martin, 2013; Sinclair & Monk, 2004).

Psychotherapeutic discourse is one of the discourses in which the clients and the therapist are already positioned in their interaction. From her/his institutional position, the therapist has the right and the duty to help the clients find out how cultural discourses work in their relationship. In the therapy of Victoria and Alfonso this help was provided by the therapists as they were actively taking part in the construction and reconstruction of the dominant stories. However, the institutional and personal position of the therapist limits her/his ability to be of help in this search (Hare-Mustin, 1994). The therapist brings into the therapy her/his cultural discourses and the psychotherapeutic discourses s/he has learnt through the psychotherapy frame that s/he has adopted. Thus, psychotherapy has also been criticized for reconstructing western cultural discourses of personhood (Guilfoyle, 2002).

The cultural discourses referred to in Victoria and Alfonso's therapy were in line with the idea of psychotherapy as a talking cure and the dominant discourse about the values attached to the romantic relationship in north-western society. Therefore, these stories had strong potential validation for emerging as dominant in couple therapy in a Northern European country. Hare-Mustin (1994) emphasized the importance of the therapist's reflexive awareness of the dominant cultural discourse as well as of the marginalized discourses that could help clients construct an alternative dominant story. For example, gendered discourses may be present in psychotherapy (Dickerson, 2013; Päivinen & Holma, 2012), and may have an effect on the therapy process if left unacknowledged.

The therapist's awareness of the marginalized and silenced cultural discourses is central in cross-cultural couple therapy, especially when the therapist shares a cultural background with one of the partners. In our results, the second dominant story also required that Alfonso changed his behavior in line with the cultural discourse that Victoria subscribed to and which is dominant in the country in which the couple live. However, acceptance of the story dominant in Victoria's home country seemed to prevent the couple from arguing and to solve the problems that lead them to seek couple therapy. Hare-Mustin (1994) suggests that how a story functions may serve as a criterion for which stories to support. In this case, these accepted cultural discourses seemed to work for this couple, since acting in line with them increased mutual understanding between the partners and eased the problems that had been constructed.

When studying power and positioning in psychotherapy, there is yet another level of positioning to be addressed—that of researcher (Parker, 2013). The present study was conducted by a female doctoral student and an experienced male researcher from a north European country. It may be argued that their reading of the data is limited to the cultural discourses of these researchers, despite their objective of reflexivity.

Conclusions

In psychotherapy and in love relationships, what constitutes normality and a good life refers to larger cultural discourses. Psychotherapy and therapists also refer to particular discourses that may structure what stories are accepted as dominant in therapy. The power of these normalizing truths may leave other possible discourses marginalized. In couple therapy, it is essential that both clients are able to narrate their experiences and that the therapist accepts these stories. Reflexivity and acknowledgement of power issues are required of the therapist in order for the therapist to be able to bring alternative, silenced, and marginalized stories into the therapy conversation. This is especially important in intercultural couple therapy, where cultural differences in discourse may be present.

Narrative and discourse analyses have been shown to be applicable to process research in psychotherapy (Avdi & Georgaca, 2007a, 2007b). These analyses are also of value to clinicians in increasing both their awareness of the power of cultural discourses in their clients' stories and their reflexivity on their practice and institutional power position in the therapy setting.

References

Avdi, E., & Georgaca, E. (2007a). Discourse analysis and psychotherapy: A critical review. *European Journal of Psychotherapy and Counselling, 9*, 157–176.

Avdi, E., & Georgaca, E. (2007b). Narrative research in psychotherapy: A critical review. *Psychology and Psychotherapy: Theory, Research and Practice, 80*, 407–419.

Bruner, E. M. (1986). Ethnography as narrative. In V. Turner & E. Bruner (Eds.), *The anthropology of experience* (pp. 139–155). Chicago: University of Illinois Press.

Bruner, J. (1987). Life as narrative. *Social Research, 54*, 11–32.

Bruner, J. (1991). The narrative construction of reality. *Critical Inquiry, 18*, 1–21.

Davies, B., & Harré, R. (1990). Positioning: The discursive production of selves. *Journal of the Theory of Social Behaviour, 20*, 43–63.

Dickerson, V. (2013). Patriarchy, power, and privilege: A narrative poststructural view of work with couples. *Family Process, 52*, 102–114.

Dickerson, V. C., & Crocket, K. (2010). "El Tigre, El Tigre": A story of narrative practice. In A. Gurman (Ed.), *A casebook of couple therapy* (pp. 153–180). New York: Guilford Press.

Foucault, M. (1977). *Discipline and punish: The birth of the prison*. London: Penguin Books.

Frosh, S., Burck, C., Strickland-Clark, L., & Morgan, K. (1996). Engaging with change: A process study of family therapy. *Journal of Family Therapy, 18*, 141–161.

Guilfoyle, M. (2001). Problematising psychotherapy: The discursive production of a bulimic. *Culture and Psychology, 7*, 151–179.

Guilfoyle, M. (2002). Rhetorical processes in therapy: The bias for self-containment. *Journal of Family Therapy, 24*, 298–316.

Hare-Mustin, R. T. (1994). Discourses in the mirrored room: A postmodern analysis of therapy. *Family Process, 33*, 19–35.

Harré, R., & Moghaddam, F. (2003). Introduction: The self and others in traditional psychology and in positioning theory. In R. Harré & F. Moghaddam (Eds.), *The self and others: Positioning individuals and groups in personal, political, and cultural contexts* (pp. 1–11). Westport, CT: Praeger.

Harré, R., & van Langenhove, L. (1991). Varieties of positioning. *Journal of the Theory of Social Behaviour, 21*, 393–407.

Hollway, W. (1984). Gender difference and the production of subjectivity. In J. Henriques, W. Hollway, C. Urwin, C. Venn, & V. Walkerdine (Eds.), *Changing the subject: Psychology, social regulation and subjectivity* (pp. 227–263). London: Routledge.

Knudson-Martin, C. (2013). Why power matters: Creating a foundation of mutual support in couple relationships. *Family Process, 52*, 5–18.

Knudson-Martin, C., & Huenergardt, D. (2010). A socio-emotional approach to couple therapy: Linking social context and couple interaction. *Family Process, 49*, 369–384.

Kurri, K., & Wahlström, J. (2003). Negotiating clienthood and moral order of a relationship in couple therapy. In C. Hall, K. Juhila, N. Parton, & T. Pösö (Eds.), *Constructing clienthood in social work and human services* (pp. 62–79). London: Jessica Kingsley.

Linehan, C., & McCarthy, J. (2000). Positioning in practice: Understanding participation in the social world. *Journal for the Theory of Social Behavior, 30*, 435–453.

McLeod, J. (2004). Social construction, narrative, and psychotherapy. In L. E. Angus & J. McLeod (Eds.), *The handbook of narrative and psychotherapy: Practice, theory, and research* (pp. 351–365). London: Sage.

McLeod, J., & Lynch, G. (2000). "This is our life": Strong evaluation in psychotherapy narrative. *European Journal of Psychotherapy, Counseling and Health, 3*, 389–406.

Päivinen, H., & Holma, J. (2012). Positions constructed for a female therapist in male batterers' treatment group. *Journal of Feminist Family Therapy, 24*, 52–74.

Parker, I. (1997). *Psychoanalytic culture: Psychoanalytic discourse in western society.* London: Sage.

Parker, I. (1998). Constructing and deconstructing psychotherapeutic discourse. *The European Journal of Psychotherapy, Counselling and Health, 1*, 65–78.

Parker, I. (2013). Discourse analysis: Dimensions of critique in psychology. *Qualitative Research in Psychology, 10*, 223–239.

Parrott, W. G. (2003). Positioning and the emotions. In R. Harré & F. Moghaddam (Eds.), *The self and others: Positioning individuals and groups in personal, political, and cultural contexts* (pp. 29–43). Westport, CT: Praeger.

Partanen, T., & Wahlström, J. (2003). The dilemma of victim positioning in group therapy for male perpetrators of domestic violence. In C. Hall, K. Juhila, N. Parton, & T. Pösö (Eds.), *Constructing clienthood in social work and human services. Interaction, identities and practices* (pp. 129–144). London: Jessica Kingsley.

Potter, J. (2004). Discourse analysis. In M. Hardy & A. Bryman (Eds.), *Handbook of data analysis* (pp. 607–624). London: Sage.

Rosen, L. V., & Lang, C. (2005). Narrative therapy with couples: Promoting liberation from constraining influences. In M. Harway (Ed.), *Handbook of couples therapy* (pp. 157–178). Hoboken, NJ: Wiley.

Sarbin, T. R. (1986). The narrative as a root metaphor for psychology. In T. R. Sarbin (Ed.), *Narrative psychology. The storied nature of human conduct* (pp. 3–21). New York: Praeger.

Silverstein, R., Buxbaum Bass, L., Tuttle, A., Knudson-Martin, C., & Huenergardt, D. (2006). What does it mean to be relational? A framework for assessment and practice. *Family Process, 45*, 391–405.

Sinclair, S. L. (2007). Back in the mirrored room: The enduring relevance of discursive practice. *Journal of Family Therapy, 29*, 147–168.

Sinclair, S. L., & Monk, G. (2004). Moving beyond the blame game: Toward as discursive approach to negotiating conflict within couple relationships. *Journal of Marital and Family Therapy, 30*, 335–347.

Sween, E. (2003). Accessing the rest-of-the-story in couples therapy. *The Family Journal: Counseling and Therapy for Couples and Families, 11*, 61–67.

White, M., & Epston, D. (1990). *Narrative means to therapeutic ends.* New York: Norton.

Winslade, J. M. (2005). Utilising discursive positioning in counselling. *British Journal of Guidance & Counselling, 33*, 351–364.

Zimmerman, J. L., & Dickerson, V. C. (1993). Separating couples from restraining patterns and the relationship discourse that supports them. *Journal of Marital and Family Therapy, 19*, 403–413.

Chapter 8
Latent Meaning Structures in Couple Relations: Introducing Objective Hermeneutics into Systemic Therapy Research

Maria Borcsa

In this chapter, my aim is to introduce a qualitative research method named Objective Hermeneutics (OH, also: Structural Hermeneutics; Flick, 2009; Flitner & Wernet, 2002) into psychotherapy research. First, an outline of my theoretical background is given. Then the main principles of the method are presented. By carrying out a step-by-step sequential analysis of a short interaction between Victoria and Alfonso during the first therapy session, the application of Objective Hermeneutics in systemic research is shown. This microanalysis is subsequently embedded into the therapy process. The chapter ends with a discussion and conclusions for couple and family therapy practice and research.

Structural Coupling, Couples, and Socialization

Following Luhmann's (1995, 2012) systems theory, social systems are constituted via communication. Using core concepts developed by Maturana and Varela (1980), Luhmann's sociology does not refer solely to social systems, but distinguishes them from biological and psychic systems, each creating an environment for the others. These systems are operationally closed and follow their autopoietical processes: a biological system functions through its biochemical operations, a psychic system through the operations of consciousness, and a social system via communication. On the one hand, all systems occur in the difference to their environments. On the other hand, they are also in an essential relation to the latter, as all systems need

M. Borcsa (✉)
Institute of Social Medicine, Rehabilitation Sciences and Healthcare Research,
University of Applied Sciences Nordhausen, Nordhausen, Germany
e-mail: borcsa@hs-nordhausen.de

© Springer International Publishing Switzerland 2016
M. Borcsa, P. Rober (eds.), *Research Perspectives in Couple Therapy*,
DOI 10.1007/978-3-319-23306-2_8

irritations from their environments in order to endure. The irritations do not function in a causal way: autopoietic systems provide in their structures a receptivity, which makes them sensitive for certain irritations—and not for others. The fitting together of structural patterns with perturbations can lead to structural coupling, prompting in the system an irritation dialectically coming from inside and outside. Therefore, structural coupling is the process which ensures that, e.g., the brain, or consciousness, or the social system of communication is constantly supplied with evolutionary irritations. For example, language serves as a coupling mechanism between consciousness (psychic system) and communication (social system). Bearing in mind the autopoietical character of each system, we can see its structures select what effect a perturbation, say, a certain sentence, may have. The term "coupling" indicates that we do not speak about one-sided causality which the environment exerts on the system but about co-evolving processes.

This formalistic sociological theory corresponds in its basic assumptions with bio-psycho-social therapy models (Borrell-Carrió, Suchman, & Epstein, 2004; Engel, 1980) and helps us to recall the complexity we have to deal with in systemic therapeutic settings taking place as communication. If what we are after is couple therapy, we have to take into account the biological and psychological backgrounds of the two persons forming a couple, which is neglected if we conceptualize couples solely as communication. To put it vice versa: even if an end of communication may terminate their *social* system, the two persons will endure on the psychological and biological levels.

As Berger and Kellner (1964) showed earlier, we have to think about a couple in terms of a "nomos-building instrumentality," where during the couple's life span two biographical backgrounds of meaning (have to) merge to create a third condition. This perspective stems from microsociology of knowledge, based on Weber, Mead, Schütz, and Merleau-Ponty, an approach where the concept of socialization plays an important role. Being socialized as a human being means to be an embodied part of a meaningful social world, while agents of socialization are all significant others who are physically and/or mentally close to a person. In the processes of lifelong socialization the symbolic organization of the world gets internalized (not necessarily in a conscious way), though it is simultaneously modified by the persons living in it (Berger & Luckmann, 1966). These two-faceted processes are biographically cumulative and transformative. A chosen life-partner may play an outstanding role in this development, being potentially an influential agent of lifelong socialization.

In this chapter, I follow a concept of lifelong socialization as structural coupling of the biological, psychological, and social systems, applying this to the phenomenon of a couple. Through communication and physical closeness a couple provides for the psychic system of the partner an environment which is likely to generate "perturbations" and which may serve as an invitation to structural coupling, including biological procreation and psychological growth. This co-evolution will create couple(d) patterns of communication and behavior, those affecting the cognitions and physical constitution of the two individuals as well.

Objective Hermeneutics as Research Method

Objective Hermeneutics (OH) is one of the most prominent approaches to qualitative research in German-speaking countries (Reichertz, 2004). The methodology was developed by Oevermann, Allert, Konau, and Krambeck (1979) (see also Oevermann, 1993) to study socialization of children in families, with the aim to investigate how socialization is accomplished. This is not self-evident, as the question is: "How can children participate in the social world of the family even though they first have to acquire the necessary competences for this?" (Titschler, Meyer, Wodak, & Vetter, 2000, p. 198). Focusing on this inquiry, the OH methodology retains the dialectics of a pre-existing social world and individual autonomy, conceptualizing socialization as an interactive process. Meaning is understood as a social structure that objectifies in concrete interaction, which can be theorized sensu Luhmann in the structural coupling of living, psychic, and social systems. Latent meaning structures precede historically and biographically the intentions of individuals and are comparable with the notion of language as a coupling mechanism in Luhmann's systems theory (for a closer discussion on the compatibility of OH and Luhmann's systems theory see Schneider, 1995).

OH can be characterized as a reconstructive procedure, aiming at latent structures and social "subconscious": it refers to those parts of meaning structures that persons are not aware of, even if those structures have factual consequences for them. In contrast to other hermeneutical methods OH transcends the intentions of individuals and therefore claims to be "objective" rather than "subjective" (for early critique of the term see, e.g., Reichertz, 1988). A methodical closeness to discourse analysis (DA) and conversation analysis (CA) can be stated (see Maiwald, 2005), as sequential analytical procedures are a shared way of interpretation. While these analytical methods are not necessarily interested in the specificity of a certain case but more in the production of communicative patterns, OH's aim is to reconstruct the specificity of a case *as well as* general patterns of social practice. For example, a certain mother–daughter interaction not only informs us about this particular relationship but also about general structures of family interaction; a single case can be seen on its own, but always represents a way of coping with general action problems as well.

In a sequential analytical procedure the hermeneutic circle (Gadamer, 1960) is applied, meaning that "all understanding is conditioned by the prior knowledge of the interpreter and that is extended through interpretation and thereby creates new conditions for understanding" (Titschler et al., 2000, ibd.). For this circle, Objective Hermeneutics offers us clear methodological concepts, research principles, and process methods, even if there are variations in the application of OH. Research material should be naturally occurring texts, keeping in mind that the term "text" is understood in a very broad sense, including social phenomena such as pictures and paintings, films or even architecture (Schmidtke, 2006).

No exhaustive English introduction to the method of OH (for an overview see Flick, 2009) exists to date. Therefore, I will follow in central aspects the German monograph by Wernet (2006) and my review of research and methodological literature in English (Flick, 2005; Hitzler, 2005; Lueger, Sandner, Meyer, & Hammerschmid, 2005; Maiwald, 2005; Reichertz, 2004; Steiner & Pichler, 2009; Titschler et al., 2000; Wagner, Lukassen, & Mahlendorf, 2010). Starting with methodological axioms, I will later address the principles pertaining to the research process and concrete research steps.

Methodological Axioms

To understand the research process of OH we need, first and foremost, some clarifications about the basic ideas of the method. We should remember the initial background of the model, namely the research on socialization. Socialization cannot be conceptualized as causality (say, as an intrusion from outside a system into a system), but more as structural coupling of different systems (biological, psychic, social). Therefore, meaning can be considered as an outcome of interaction, e.g., as meaning for the participants themselves, but also as independent of the participants, being part of the system of communication. Let us have a closer look at these axioms and the implications they have for research.

Text interpretation Oevermann's central assumption is that all social science has to be considered as an interpretive science, as our social world is meaningful ("sinnhaft") and all empirically verifiable statements about the social world are interpretations. OH focuses on the methodological monitoring of the scientific operation of interpretation.

Text as a rule-generated configuration Social interaction does not develop randomly but is based on rules. Every social practice is framed by a rule-governed space, providing possibilities for acting, e.g., constitutive language rules provide participants of communication with a certain range of possibilities as to how to use an utterance. Society's members have tacit knowledge or rule competence, which they normally acquire during socialization. This rule competence functions in concrete social practice and it is also used in the process of OH analysis as the interpreter(s) use their own rule competence for interpretation. Therefore, the concept of rules is to be seen as a bridge between a certain social practice and the method of OH.

Structures of meaning The participants of an interaction do not have complete authorship of the meaning structures in the interaction: they act with a certain autonomy in the sequentially developing interaction structure by choosing specific interactive options over others. Constitutive language rules provide us with a frame of possibilities to communicate, whereas a concrete interaction is *one realized* possibility (OH shares this assumption with CA and DA). Consequently, OH distinguishes between the latent and the manifest structure of meaning. While the latent meaning structure refers to this field of given possibilities of which the actors are normally not fully conscious, the manifest meaning structure refers to the subjective representation. The aim of the research is not to try to get as close as possible to the subjective perspective of the actors, but to explicate the *differences* between the latent and the manifest structure of meaning. We will come back to this later.

Case structure To understand what a case in terms of OH means, we could use an example referring directly to systemic therapy. Couple therapy can be considered as a social practice where different levels come into play: the individual biographical level (with the respective personal socialization effects by agents like family of origin, institutions, political structures, etc.), the level of the couple as a social system, and, finally, the therapeutic system, comprising the couple and the therapist(s). Each of these levels can become the focus of analysis in terms of OH. Based on the possibilities the latent structures of meaning provide on each level, social actors (be it an individual, a couple, or a therapeutic system) choose from these options. These choices are not contingent, but follow a certain pattern. The latent structures offer different possibilities to (re-)act, but *how* a system (person, couple, therapeutic system) is actually acting gives us information about the structure of the case under investigation.

Case structure reconstruction through sequential analysis The selectivity of a case described above is not static, but has to be conceptualized as a process. The case analysis in the logic of OH does not collect and systematize data. Rather, the researcher reconstructs the sequence structure of the social practice. From a methodological point of view, she/he focuses on the aspect of structural reproduction in the research process, taking into consideration that the case is not completely "inventing itself" on the spot. Instead, a case is considered to be based on dynamic structures (Oevermann is following here Mead, 1934, especially Mead's distinction of *I* and *Me*), which the researcher reconstructs in the analysis.

One assumption is, therefore, that the case reconstruction can be done with any sequence the material provides - even if there are differences in the richness of the material. For example, if we wish to analyze the therapeutic system, it is useful to start with the very beginning of the conversation in the first session, as we are able to follow how the system is forming its patterns and structure (see Chaps. 3 and 4 in this book). If a communication system has a longer history—like that of a couple— our decision of choosing the utterance to be analyzed may follow other criteria. I will describe this in the next paragraph.

The Research Process: Analysis of a Sequence by Victoria and Alfonso

The theoretical key points are rather abstract and can leave OH novices at a loss. In order to clarify the principles and phases of the research process (Titschler et al., 2000; see also Wagner et al., 2010; Wernet, 2006) I present a step-by-step analysis which serves the purpose of getting acquainted with OH.

First, the researcher should *define the aim of research as precisely as possible. What is the "case" in our analysis?*

The research aim in this study is to *reconstruct the dynamic pattern of the couple*, which might relate to bringing them into therapy and to examining if and, respectively, how this pattern might undergo change. So I will not analyze the therapist(s) and their interventions but focus on the couple as a system.

In the analysis I assume that the couple is not inventing an entirely new "dance" in front of the therapist(s), but performs patterns of the couple dynamics in the session. Yet, we should not forget that they do not cease to be individuals as well (with the potential autonomy to end the couple system), and their biographically socialized individual patterns are also brought into interaction. The therapists as conversational partners fulfill a professional task, never totally free from representing also institutional configurations (see Peräkylä, Antaki, Vehviläinen, & Leudar, 2008 and Chap. 7 in this book), which frame the therapeutic system by communicative procedures. The therapists have to be considered as socializing agents themselves.

As the second step, a decision has to be taken *on the material to be used in the first sequential analysis according to the case of interest.*

The sequence for analysis (see below) was chosen for reasons of being an instance of interaction between the partners dealing with a central and emotionally loaded subject without the therapist intervening in their exchange. We can look at this sequence as a spontaneous enactment (Minuchin, 1974) of the couple's dynamics. The sequence analyzed takes place 1 h into the first therapy session (total time of this session: 1 h 22 min).

I will now turn directly to this sequence by presenting the research principles and method of OH. The principles of OH explained and applied are (1) verbatim approach, (2) extensivity, (3) freedom of context, and (4) sequentiality (Wernet 2006).

First principle: verbatim approach OH's methodological premise that our social world is meaningful ("sinnhaft") and all social practice can be considered text-like has its methodological equivalent in the consistent literalness of analysis.

Second principle: extensivity OH can be considered as a microanalytical approach, going very much into detail of utterances and interactions. This has to be regarded methodologically against the background of the axiom that it is possible to abstract general structures from a special case. As a result, OH uses small amounts of research material, following the claim not to describe or paraphrase but to reconstruct the generative rule (which explains the pattern) of the case.

Third principle: freedom of context During analysis, *the first chosen utterance* will be looked at without taking into account any contextual knowledge. This is a methodological decision to approach the latent meaning structures of a certain utterance or interaction. A context-free explication of an utterance uses *the method of thought experiment* by originating contexts which are compatible with the utterance under investigation. In doing so, the interpreter avoids reconstructing the meaning of the utterance solely from the actual context of the utterance. Bringing in the actual context is a methodically controlled step.

Fourth principle: sequentiality This principle has a central position in OH: the analyzer has to take the sequential construction of meaning into account; that is, she or he should not interpret, say, a question by using the following answer. Only by a strict application of this standard can we reconstruct the structural logic of a certain interaction. The so-called inner context of a case develops again through thought experiments: *what can happen next?* (e.g., after a certain question: who could speak, what could this person say, what arguments can be expected?) By comparing the real continuation with the previously generated possibilities we can develop hypotheses about the case structure under consideration.

Let's now introduce the first utterance of the analyzed sequence:

Speaker 1 (S1): *"They own you"*

Following the method of thought experiment we ask ourselves: In which context could this utterance occur in a meaningful way?

Analysis (1) In the *literal sense* of this utterance there is only one possible meaningful context: that of slavery/human trafficking. Here we could have a more powerfully positioned person saying this to a less empowered one or a representative of the older generation to a younger one, e.g., as an admonishment. (2) In a *metaphorical* way there seem to be two possible interpretations: (a) "they" could refer to relational contexts like family or friends having massive emotional power over the object of the sentence ("you") or to an institutional/organizational system like the working place exercising its total structural authority over the object (i.e. object has no rights); (b) "they" does not refer to persons but to certain powerful inanimate objects like drugs or hallucinatory voices getting out of control.

Now we look for common structural aspects of the created contexts. Leading question: What are the structural features of these types of contexts, especially in terms of positions of the interlocutors?

Analysis The utterer labels, delivers a verdict on the other person's condition: S1 gives a statement from an observer position, while this sentence may express concern, compassion but also superiority or a certain kind of triumph. The utterance places the object "you" into a maximally subjugated position concerning the subject "they" (the hegemony of "owning" implies an inhuman totality). Literally or figuratively, "you" is placed under control in a situation of constraint—and this may be understood by the listener/addressee as a request, demand, challenge, or even as a provocation for acting.

Outline possible options for further interaction What options are available for the next meaning unit? *The leading question is: What could happen next?*

1. *S1 continues*, e.g., gives an explication for this statement or tells something about the background of this judgment, e.g., in a narration.
2. *Someone else reacts, the "you"*: s/he (a) could agree; (b) could ask for explanation; (c) could remonstrate overtly; (d) could remonstrate covertly; (e) could change the topic.
3. *Someone else reacts, who is not the "you"*, e.g., asking for explanation or coming up with a different topic, etc.

Second and following utterances of the selected sequence:

We now compare the next meaning unit (utterance) with the available options by characterizing the linguistic features of the unit at the syntactic, semantic, and pragmatic levels. *Leading questions: What are the specifics of the following realized utterance? What grammatical form was used (active, passive, conditional, etc.)? What themes are mentioned? Are there any linguistic peculiarities (slips, breakdowns, use and misuse of words)?* Then we explicate the function of the unit in the distribution of interactive roles and positions. *What relations and attributions to persons are given (even if not directly named) or could be implied in the text? (Titschler et al. 2000, p. 206).*

Contrast with the actually chosen turn:

Speaker 2 (S2): *"They think they own me [laughs]"*

Analysis This utterance subjectivizes the declared fact of "owning" into a cognition (refers to "their" subjective construct of reality), while the speaker does not deny the entire purport of the previous account. The change in meaning proposed is the weakening of totality through this subjectivized construction "They think they..." (which excludes context (2b), as inanimate objects can't "think") and, through that, a freedom regained by the object of the utterance ("me"). The "owning" becomes one-sided through this shift, a 'unilateral agreement' as a special way of relating, *demonstrating that S2 is independent or able to break free from this constellation whenever s/he wants.* By selecting this verbal possibility (S2 chooses (d) out of the possible reactions outlined above) and by, for instance, not asking S1 for a clarification of the statement, S2 takes the topic as established, shared, and significant in the interaction (S2 is not changing the topic). The laughing might refer to a position of felt superiority and/or unease in speaking about this topic, or may serve as attenuation. This (covert) remonstration makes the first utterance evolve into an argumentative statement, whereupon S2 parries with a counter-argument. This counter-argument may function in that it plays down the content of S1's utterance—for all listeners and/or for the speaker him/herself.

Summing it up, S2 may follow a "hidden agenda" in the relationship with "them," e.g., for conflict-avoiding and/or benefit-seeking reasons.

Proceed with the same operations: explicate the potential action space opened by the meaning-structures and contrast it with the actually chosen options.

1. S2 continues to speak (a) staying with the topic by, e.g., giving an explanation for the counter-statement; or (b) changing the topic
2. S1 takes the turn (a) asking for an explanation; (b) remonstrating overtly; (c) remonstrating covertly; or (d) changing the topic
3. S1 laughs as well (shares the meaning implicitly)
4. Someone else joins the conversation

Contrast with the actually chosen turn:

Speaker 1: *"Yeah but as long as they own you they have a right to control you"*

Analysis Although s/he gives an assertive marker ("yeah") the first speaker sticks to the original assumption, pursuing her/his own line of argument, at first sight without referring to the preceding utterance of S2. Only the word "but" seems to relate this sentence to the previous one, announcing an objection. On the content level, S1 is not referring to the *subjectivization* conveyed by S2, but is explicitly pointing out "control" as a structural aspect of ownership. Moreover, s/he introduces a new facet of "owning": having the "right" to do something with the object in question. S1 presents—while ignoring S2's remonstration and her/his endeavor to point out her/his liberty—a frame of reference distinct from that of S2, entailing an objective order of rights, an indisputable verity, thus an *objectivization*. S/he frames this by means of a temporal marker "as long as," denoting not only a status but also a potential change in due time.

From an interactional perspective, this semantic and pragmalinguistic choice forms a new round of the argumentation. S1 is speaking again from the position of a noninvolved observer, giving once again an assessment of the other person's situation. Speaker 1 monologizes through this extended repetition without picking the thread of the other speaker's previous utterance, demonstrating to the listener(s) and to her-/himself that s/he seems, at present, more convinced by her/his own representation (of hegemony) than by the words and the idea (of freedom) of the other person.

Continuation of the analysis Potential action space opened by the meaning structures:

1. S1 continues to speak (a) staying with the topic by, e.g., giving an explanation for the counter-statement; or (b) changing the topic
2. S2 takes the turn (a) asking for an explanation; or (b) remonstrating overtly; (c) remonstrating covertly; or (d) changing the topic
3. Someone else joins the conversation

Contrast with the actually chosen turn:

Speaker 1 (continues) *"and like really there's this strong, pressuring thing I feel"*

Analysis S1 continues to speak, now including him/herself into the frame of reference. While "like" may show a search for words, hesitation or a difficulty to start, the term "really" validates the following utterance in advance. The connection to the previous sentence on the content level is not evident, whereas S1 comes up with the vague expression "thing," specified as "strong" and "pressuring." This object or entity is referred to as inducing an emotional state in S1, while it stays unclear if "there" refers to the existence of this entity generically, or only circumstantially (at a certain time or a certain place). As S1 speaks of *"this"* and not *"a"* "thing," the entity is precised and accentuated. This emphasis is augmented through the word order, as the speaker comes up with her/his own perception of the vague entity—not as an observation or a cognition but as a sensation—at the end of the sentence and not at the beginning. Though the speaker implicitly creates a self-description as a sensitive person, the "thing" itself seems to be accountable for her/his emotional state through its massiveness. What may this choice of continuation tell us? Is S1 staying with the topic or changing it? The connection between the two utterances could be a backing of the counter-statement. If we follow this hypothesis, then emotions are used here as a validating authority.

Potential action space opened by the meaning structures:

1. S1 continues to speak, giving an explanation of the inexplicit declaration
2. S2 takes the turn,

   - going back to the first utterance in this sequence of S1 (a) asking for an explanation; (b) remonstrating overtly; or (c) remonstrating covertly
   - staying with the second utterance in this sequence of S2 (a) asking for an explanation of the inexplicit declaration; (b) remonstrating overtly by denying existence of the "thing" or the "feelings" of S1; or (c) remonstrating covertly
   - changing the topic

3. Someone else joins the conversation

Contrast with the actually chosen turn:

Speaker 2: *"But for example I don't, how'd you say, I don't care about that, I just don't care"*

Analysis S2 also connects using the word "but," marking an objection and a continuation of the dispute. While starting with "for example," which would bring in a special situation or case, S2 interrupts him-/herself. The fillers "I don't, how'd you say" show now a search for appropriate wording on her/his part. S2 (again) does not disclaim overtly the content of the preceding account but expresses a personal attitude to a phenomenon correspondingly vaguely described with "that." S2 behaves as if there is a (silent) agreement between S1 and S2 on the respective content of their utterances, as ambiguous phenomena are not explicated further and not questioned either. Now, too, S2 follows the line of her/his own argument of freedom and independence here by referring to a personal standpoint (of not caring) as regards the "rights" others may have, also emphasizing it by a repetition.

Interestingly, as "that" is not concretized, it can also refer to the second part of S1's utterance: to the feelings S1 is mentioning. The "not caring" might refer to the "strong, pressuring thing" and thus mark a contrasting stance one may assume toward "that." However, it may also be interpreted as S2's attitude toward the feelings of S1, and thus being understood as an open refusal to empathize.

Again, this echoes an emotionally detached and autonomous position, not only toward the structural aspect of control mentioned, but also against the emotional charge voiced by S1 in the present interaction.

What could happen next?

1. S2 continues to speak, (a) giving an explanation for this statement; (b) now starting with a (provocative) statement toward S1 him/herself; or (c) introducing a new topic, thus declaring the dispute terminated
2. S1 takes the turn, (a) asking for an explanation for the "not caring" part of the statement; (b) pursuing his/her own line of argument (repeating the previous

interactional pattern); or (c) introducing a new topic, thus declaring the dispute terminated

3. Someone else takes the turn

Contrast with the actually chosen turn:

Speaker 1: *"I just do"*

Analysis The shift of S1 from an external to an internal position in the conversation is completed now: the subject of this sentence is the speaker her/himself, while S1 places her/himself in a binary counterposition to S2, which creates a confrontation. The validity of the "not-caring" proposal as a possible option is not denied, but emphatically (the term "just" is borrowed from S2) not accepted as a choice for S1, interactionally reinforcing the statement s/he gave before. We may say that S1 is again pursuing his/her own line of argument, with the outcome that the repetition of the diverse standpoints creates a communicative escalation and a dead end.

Summary of the Analysis by Introducing the Case

We are dealing with a couple therapy conversation: S1 and S2 form the couple, speaker 1 being female (Victoria), speaker 2 being male (Alfonso). The communication system consists of the two persons speaking with each other and a third person who is not involved here in the actual sequence but is a listener (a male therapist). The concrete verbal context is that the couple speaks about Alfonso's family of origin (see above context variation: context 2a).

348V	They own you
349A	They think they own me [laughs]
350V	Yeah but as long as they own you they have a right to control you and, like really there's this strong, pressuring thing I feel
351A	But for example I don't, how'd you say, I don't care about that, I just don't care
352V	I just do

Taking this sequence as an observer's punctuation (Watzlawick, Beavin Bavelas, & Jackson, 1967) in a dynamics of this dyad, we may say that Victoria's sentence (348V) introduces thematic tension, whereas Alfonso stays in verbal contact without explicitly agreeing or disagreeing but coming up with a third idea (of personal sovereignty, 349A). Interactional pressure in the next utterance of Victoria (350V) is increased through her reproduction of the content and the use of appeal to emotions and feelings as validation of credibility; we can describe Victoria's rhetorical style as emotive; fillers-cum-intensifiers add emphasis to the content of the statement. Alfonso's reaction (351A) is structurally similar to the one before (remonstrating covertly), referring now explicitly to his independence through emotional detachment. The sequence creates a situation where Victoria is following her own agenda/schema without taking into consideration the dissimilar meaning proposals of Alfonso, whereas Alfonso reiterates to position himself as self-regulated and (emotionally) independent. This seems to give Victoria a new impulse to reproduce her own communicative pattern.

We may say that the *interactional* behavior (pressure through emotionalized verbal behavior) shows an isomorphic structure (Bertalanffy, 1968) to the *content* ("pressuring thing") of Victoria's speech, while Alfonso shows a refusal to take on either the content or the appeal of Victoria's verbal behavior, correspondingly performing what he is saying. In the course of the argumentation neither the issue nor the tension are resolved; moreover, the latter escalates and ends up in a confrontation.

Development of Structural Hypotheses about the Case under Investigation

By extrapolation of the interpretation of the unit onto recurrent communicative patterns we will look now at relational aspects that may transcend the situation. Doing this, we can expose general relations and structures.

Generalizing hypotheses The two persons perform on the content level as well as on the interactional level the dialectics of emotional involvement vs. detachment. This configuration may create, like in the analyzed sequence, an escalation in concrete communication. Furthermore, in the long run, a negative circle could develop over time as a result of repeating this pattern—on a relational level these oppositions might end up in a demanding (Victoria) vs. a withdrawing (Alfonso) position. We hypothesize that the "choice" of these positions is not coincidental but is based on biographical configurations grounded also in gender structures offered by society's socializing agents.

Until now we have been expounding the material without taking into account information about the external context. The aim was to inductively analyze the communication pattern without going into an explanation using facts which are not in the material itself. As the next step, we will introduce the external context.

Introducing the Therapy Process

Not to repeat the background of the therapy here (please see Chap. 2), it should be remembered that Victoria had followed a course of individual therapy before starting couple therapy with her partner. As she feels much better now, the goal is to get some help with their "communication problem," as Victoria summarizes their disputes (first session 390V). Following our generalizing hypothesis, we assume that the analyzed sequence shows both personal as well as relationship patterns and gives some insight into the couple's problematic dynamics. We will expound this in more detail by taking into consideration the therapy process now.

Victoria, who has infrequent contact to her own separated parents, experienced a rejection in Alfonso's family of origin already during the first visit of the couple to his home country. In Victoria's (and her individual therapist's) narration this played an important part as a trigger for her becoming depressed, introduced very early in the first therapy session. Victoria lives with a lack of control over what is happening between Alfonso's family of origin and her partner. She is excluded from their communication in a double sense: being physically in their presence she cannot understand and speak the language. But also the decision not to visit Alfonso's family because of the experienced rejection is only partly a good solution for her, as she does not know what is going on when Alfonso is in his home country. In the second session she says V243: "Yeah (.) and yeah it is a big thing because I know that family is important and I also (.) I have tried also to, I don't want Alfonso to be between two families," 244T2: "mm" 245T1: "mm," 246V: "I think I have tried but then I feel do they want to keep you so busy so that (.) you don't have time for me, because it's clear that they don't like me." Her valuation in the same session is clear: 529V: "(…) they are passive aggressive and manipulative." Her hint that Alfonso could "talk" about the familial relations and, by doing so, clarify her position is not taken by him, see second session, 409A: "Yes then it would also easier that you [to V] could come there, but I feel that's it just, that it could just be me, and I don't know how to say, it's that's just the way I feel about them somehow, I just feel too forced to, now to just change that, it feels something not natural, it would be something not natural." 410T1: "To try to solve this or?" 411A: "Yeah." His decision of distancing himself (in the literal and metaphorical sense) seems to be a strategy which works for him personally, although his parents are not willing to come to visit him/them (a fact Victoria was criticizing already in the first session). He says in the second session, 296A: "But my parents, that's maybe why when (.) when I'm there also I don't spend that much time with them, I don't know, I somehow don't miss them too much, so to me it's like I've always needed my independence and" 297T1: "Mm" 298A: "Maybe somehow I see, my being here is a sort of independent thing, maybe if they don't come" 299T1: "From your parents?" 300A: "Yes" 301A: "So if they don't come here it's not a big deal to me." With regard to Alfonso's family background, the couple's circumstances generate ambiguity and diversity in meaning, which seems to challenge Victoria. Her emphatic statement about the type of relationship between Alfonso and his family (content level of the sequence examined above) and her monological style of communication (interaction level) puts into the analyzed scene a desperate and provocative call to action toward her partner for a completion of a disjuncted personal "Gestalt" (Perls, Hefferline, & Goodman, 1994) to contribute to the easing of her lack of confidence (and implicitly to show his attachment this way). By sticking to his communicative and personal pattern of positioning himself as autonomous and emotionally independent, Alfonso, on his part, calls for, proposes and models tolerating ambiguities and not giving in to strange, pressuring forces. Through holding on to this pattern, he challenges Victoria's self- and worldview; she cannot easily consent to the usefulness of emotional independence, as this would mean that she has to tolerate this personal pattern also with respect to herself as a meaning and rule creation for the couple itself (the building of a "nomos"). Her position "I am asking that I am the most important thing in his life" (second session, 74V) focuses, in contrast, on (total) affiliation and not on independence, while Alfonso's commitment to the relationship seems not that decisive; at the end of the second session he says 528A: "(…) if I really had to choose and I know you wouldn't make me choose, if they would make me choose, I would never choose them (.) do you understand?". This is a distancing position from his family of origin, but not a revelation of belonging to Victoria.

Between the second and the third session Victoria starts working less; third session 15V: "Like after this week, I started working much less (.) it means less money but it means more, more life of my own and for us also, we have time to see each other, so, I think it's for the best." Further she learns that Alfonso missed her when he was alone in his home country the weeks before, an experience that generates also new viewpoints for her; 70V: "Yeah and I think that it is just something we see differently, that (.) yeah, I don't think it's any big deal now because when he came back from [A's home country] I saw that, like he had missed me and everything was fine." Alfonso, being asked about the meaning he gives to the relationship, states later in the session, 222A: "Yeah, yes, the first time of (..) kind of strong and important, relationship" and A240: "I am like happy with that life." Showing this to Victoria makes it easier for her to tolerate differences and ambiguities.

During the fourth and last therapy meeting, the couple is looking for the reasons why they have had a good time lately. Victoria says, 83V: "And I try to think more and more like, I have these issues with trust, if there is a small thing that, like a very small thing can make me feel like abandoned and that he doesn't love me any more (.) and I try to take, think like rationally and I force myself to believe in his words even if my (.) like something inside me is telling me 'don't believe anything you hear', I force myself to believe because if I don't believe his words then I don't have anything." Victoria explains how she tries to change her personal pattern of emotional immersion. Her construction is that of a personal struggle between two qualities of herself ("I" and "something inside me"). The emotionally doubting one tries hard to transform into a rationally "convinced" one, giving up the fight for an absolute certainty and a total affiliation, which is structurally impossible. Even if Alfonso's ability for and need of emotional detachment may be a threat to Victoria (and to the couple system itself), in the long run it might also be a successful stance in freeing the couple from, making them "independent" from, i.e., resilient against depression.

Discussion

Change during the time of couple therapy can occur on different planes: on the level of personal patterns (cognitions, emotions, behavior) and/or on the couple level (the communicative and interactional patterns). Systemic theorists and therapists assume a feedback loop between these processes, with the therapist serving as an agent of perturbation in these patterns (Ludewig, 1989). The microanalysis of the Victoria and Alfonso sequence showed us that the couple performs on the content level as well as on the interactional level the dialectics of emotional involvement vs. detachment. Furthermore, we saw that this conflict is embedded in the couple's (implicit) "nomos"-building negotiation on the structural aspects of affiliation vs. autonomy every couple has to deal with. Discussing the aspect of change we may say that the couple finds in therapy a way of dealing with their so-called "communication problem" but they do not work on their generative rule of affiliation vs. autonomy. Victoria is still caught up in a personal schema of totality when she states "if I don't believe his words then I don't have anything." The implicit decision of the couple (and the therapist) to stick to the manifest and not work on the structural level might be wise at that moment: otherwise both partners would be challenged to take more explicit positions concerning their relationship—a communicative act which could also lead to the end of the couple as a social system.

The case of Victoria and Alfonso has two specific aspects at which I would finally like to have a closer look: being a couple burdened with depression and being an intercultural couple; the case analysis will be enriched with existing literature on these aspects.

Couple with Depression

Victoria and Alfonso refer to their "communication problem" as "scars" left by Victoria's depression on the couple's level and the reason to come together to therapy. Reviewing the literature on couple therapy and the treatment of depression, Beach, Dreifuss, Franklin, Kamen, and Gabriel (2008) point out that over time a negative circularity is likely to develop in burdened couples: stress is introduced into the relationship through so-called depressive ways of behaving like "excessive reassurance seeking." This elicits changes in the partner's view, creating a potentiality for a negative feedback and the establishment of a vicious cycle, which leads again to more distress in the couple. Beach et al. (2008) conclude: "As a consequence of these converging lines of research, we currently view depression and marital discord as components of a larger vicious cycle that creates a self-sustaining loop." Especially when the time together is more and more used for fights, the danger is that "the positives in relationship may erode over time" (p. 547).

The analysis of Victoria and Alfonso's case supports this viewpoint. The sequence analysis using Objective Hermeneutics highlights the manifest pattern of the so-called "communication problem," but it also reveals an underlying structural problem. The communication problem can be seen as the manifest meaning structure or the symptom of the couple's struggle with the fundamental question of affiliation vs. autonomy, wherein the positions of the partners seem to be clearly marked: Victoria seeks for (signs of) attachment, Alfonso 'needs his independence'. Alfonso, who is some years younger than Victoria, left his home country for studying,

Victoria herself was already working when they met; their ideas of the nearer and the more distant future may likely have been different already in the beginning of their relationship, building a biographical background for the development and consolidation of these positions. Dialectically at this stage of the couple's development what induces their problem is also a protection of their social system: in therapy the couple works with, and improves, their "communication problem". But the 'nomos' stays ambiguous: the (un-)shared needs, values and beliefs are not explicitly discussed, but held in abeyance.

An Intercultural Couple

Existing literature sources (Bhugra & de Silva, 2000; Bustamante, Nelson, Henriksen, & Monakes, 2011; Hsu, 2001; Molina, Estrada, & Burnett, 2004; Sullivan & Cottone, 2006) name extended family relations a primary stressor in intercultural couples: for example, partners feel excluded from their spouse's family of origin, particularly when they are not fluent in the partner family's native language. This leads to an experience of marginalization, especially when family members speak their mother tongue instead of choosing the language of the in-law or a possible third one; that often goes together with misinterpretations and suspicions or conflicts. Further problems might be misguided expectations of the partners based on projected beliefs about other cultures, differences in ways of coping, different concepts of family boundaries and obligations, and conflicts over role division between spouses (Hsu, 2001). The special task is either to find a modified middle path that suits, or at least a path that does not conflict with, either side or culture (ibid.). Bustamante et al. (2011) show that resilient intercultural couples use mechanisms of "cultural reframing." This "third culture building" (ibid.) takes the intercultural relationship as an opportunity to learn and grow in ways that one would not in a same-culture relationship. Even if this process is structurally similar to all couples' developmental task of building their own horizon of common beliefs, values, and norms, some understandings, principles, and behavior might differ in a greater way because of socialization in different cultures. On the other hand, these differences may well be attractive because they free from aspects of one's own culture of origin that are experienced as disagreeable or restraining (marrying-out) (Rosenblatt, 2009).

Regarding the case of Victoria and Alfonso, the above cited positions of affiliation vs. autonomy are complicated by the dynamics of culturally based differences. Victoria wished to be involved in Alfonso's family. The experienced rejection and the meaning given by her to these Mediterranean family relations increase her personal strain and the pressure exerted by her on Alfonso. Her own self- and worldview seems to have traces of a "protestant ethic" (Weber, 1930), with an emphasis on duties and life constructed as struggle - a perspective others and oneself are evaluated from. Interestingly, the main change in therapy seems to happen between the second and the third session, a time where Victoria started working less to have more leisure time for herself and with her partner. How much this is a trigger for—or a result of—change, we cannot verify.

Conclusion

Microanalysis of couple therapy transcripts as a means of investigating the couple (or a family) is a fruitful endeavor for a researcher as well as a practitioner: through the explication of the possibilities the latent meaning structure of language offers us, we can elaborate alternatives in communicative patterns together with our clients. This is also in tune with the idea of empowering our clients to make fuller use of their options, even if there is no escape from discourse as such.

Objective Hermeneutics can also be seen as a time- and resource-consuming endeavor especially if it is done, as recommended, in a group. Concerning psychotherapy research, the method gives us a possibility to reconstruct the level of change: does a couple or a family create a second order change by altering also their structural rules in addition to their patterns? The relevance is vibrant, as the disruption of dysfunctional interactional cycles and patterns is an important common factor throughout different couple and family therapy approaches, even with very diverse philosophical backgrounds (Sprenkle, Davis, & Lebow, 2009); furthermore, all of the best empirically validated approaches to relational therapy utilize interventions that are focused on pattern disruption (Sprenkle, 2002). Especially in couple therapy with a partner diagnosed with depression, the early disruption of a possibly evolving dysfunctional pattern presents high importance. However, the issue how depression can be conceptualized as an implicit solution for relational structural problems still needs further research.

References

Beach, S. R. H., Dreifuss, J. A., Franklin, K. J., Kamen, C., & Gabriel, B. (2008). Couple therapy and the treatment of depression. In A. S. Gurman (Ed.), *Clinical handbook of couple therapy* (4th ed., pp. 545–566). New York: Guilford.

Berger, P., & Kellner, H. (1964). Marriage and the construction of reality: An exercise in the microsociology of knowledge. *Diogenes, 12*, 1–24.

Berger, P., & Luckmann, T. (1966). *The social construction of reality. A treatise in the sociology of knowledge*. Garden City, NY: Anchor Books.

Bertalanffy, L. von (1968). *General system theory: Foundations, development, applications*. New York: Braziller.

Bhugra, D., & de Silva, P. (2000). Couple therapy across cultures. *Sexual and Relationship Therapy, 15*, 183–192.

Borrell-Carrió, F., Suchman, A. L., & Epstein, R. M. (2004). The biopsychosocial model 25 years later: Principles, practice, and scientific inquiry. *Annals of Family Medicine, 2*(6), 576–582.

Bustamante, R. M., Nelson, J. A., Henriksen, R. C., Jr., & Monakes, S. (2011). Intercultural couples: Coping with culture-related stressors. *The Family Journal, 19*(2), 154–164.

Engel, G. (1980). The clinical application of the biopsychosocial model. *The American Journal of Psychiatry, 137*, 535–544.

Flick, U. (2005). Qualitative research in sociology in Germany and the US — state of the art, differences and developments [47 paragraphs]. *Forum Qualitative Sozialforschung/Forum: Qualitative Social Research, 6*(3), Art. 23. Retrieved from http://nbn-resolving.de/ urn:nbn:de:0114-fqs0503230.

Flick, U. (2009). *An introduction to qualitative research* (4th ed.). London: Sage.

Flitner, E., & Wernet, A. (2002). Faire de la sociologie avec des étudiants. De l'usage de "l'herméneutique structurale" dans la formation des enseignants. *Education et Sociétés. Revue internationale de sociologie de l'education, 9*(1), 101–113.

Gadamer, H.-G. (1960). *Wahrheit und Methode*. Tübingen: Mohr.

Hitzler, R. (2005). The reconstruction of meaning. Notes on German interpretive sociology [35 paragraphs]. *Forum Qualitative Sozialforschung/Forum: Qualitative Social Research, 6*(3), Art. 45. Retrieved from http://nbn-resolving.de/urn:nbn:de:0114-fqs0503450.

Hsu, J. (2001). Marital therapy for intercultural couples. In W. Tseng & J. Streltzer (Eds.), *Culture and psychotherapy* (pp. 225–242). Washington, DC: American Psychiatric.

Ludewig, K. (1989). 10 + 1 guidelines or guide-questions. An outline of a systemic clinical theory. In J. Hargens (Ed.), *Systemic therapy. A European perspective* (pp. 13–32). Borgmann: Broadstairs.

Lueger, M., Sandner, K., Meyer, R., & Hammerschmid, G. (2005). Contextualizing influence activities: An objective hermeneutical approach. *Organization Studies, 26*(8), 1145–1168.

Luhmann, N. (1995). *Social systems*. Stanford, CA: Stanford University Press.

Luhmann, N. (2012). *Introduction in system's theory*. Chichester: Wiley.

Maiwald, K.-O. (2005). Competence and praxis: Sequential analysis in German sociology [46 paragraphs]. *Forum Qualitative Sozialforschung/Forum: Qualitative Social Research, 6*(3), Art. 31. Retrieved from http://nbn-resolving.de/urn:nbn:de:0114-fqs0503310.

Maturana, H., & Varela, F. (1980). *Autopoiesis and cognition: The realization of the living*. Dordrecht: D. Reidel.

Mead, G. H. (1934). *Mind, self, and society*. Chicago: University of Chicago Press.

Minuchin, S. (1974). *Families and family therapy*. London: Tavistock.

Molina, B., Estrada, D., & Burnett, J. A. (2004). Cultural communities: Challenges and opportunities in the creation of "happily ever after" stories of intercultural couplehood. *The Family Journal, 12*(2), 139–147.

Oevermann, U. (1993). Die objektive Hermeneutik als unverzichtbare methodologische Grundlage für die Analyse von Subjektivität. Zugleich eine Kritik der Tiefenhermeneutik. In T. Jung & S. Müller-Doohm (Eds.), *"Wirklichkeit" im Deutungsprozeß: Verstehen und Methoden in den Kultur- und Sozialwissenschaften* (pp. 106–189). Frankfurt/M: Suhrkamp.

Oevermann, U., Allert, T., Konau, E., & Krambeck, J. (1979). Die Methodologie einer objektiven Hermeneutik und ihre allgemeine forschungslogische Bedeutung in den Sozialwissenschaften. In H.-G. Soeffner (Ed.), *Interpretative Verfahren in den Sozial- und Textwissenschaften* (pp. 352–434). Stuttgart: Metzler.

Peräkylä, A., Antaki, C., Vehviläinen, S., & Leudar, I. (2008). Analysing psychotherapy in practice. In A. Peräkylä, C. Antaki, S. Vehviläinen, & I. Leudar (Eds.), *Conversation analysis and psychotherapy* (pp. 5–25). Cambridge: Cambridge University Press.

Perls, F., Hefferline, R., & Goodman, P. (1994, original 1951). *Gestalt therapy: Excitement and growth in the human personality*. Gouldsboro: Gestalt Journal Press.

Reichertz, J. (1988). Verstehende Soziologie ohne Subjekt? Die objektive Hermeneutik als Metaphysik der Strukturen. *Kölner Zeitschrift für Soziologie und Sozialpsychologie, 40*, 207–222.

Reichertz, J. (2004). Objective hermeneutics and hermeneutic sociology of knowledge. In U. Flick, E. V. Kardorff, & I. Steinke (Eds.), *A companion to qualitative research* (pp. 290–296). London: Sage.

Rosenblatt, P. C. (2009). A systems theory analysis of intercultural couple relationships. In T. A. Karis & K. D. Kilian (Eds.), *Intercultural couples. Exploring diversity in intimate relationships* (pp. 3–20). New York: Routledge.

Schmidtke, O. (2006). *Architektur als professionalisierte Praxis. Soziologische Fallkonstruktionen zur Professionalisierungsbedürftigkeit der Architektur (Forschungs-beiträge aus der Objektiven Hermeneutik, Bd. 8).* Frankfurt/M: Humanities Online.

Schneider, W. L. (1995). Objektive Hermeneutik als Forschungsmethode der Systemtheorie. *Zeitschrift für soziologische Theorie, 1,* 129–152.

Sprenkle, D. H. (2002). *Effectiveness research in marriage and family therapy.* Alexandria, VA: American Association for Marriage and Family Therapy.

Sprenkle, D. H., Davis, S. D., & Lebow, J. L. (2009). *Common factors in couple and family therapy. The overlooked foundation for effective practice.* New York: Guilford.

Steiner, P., & Pichler, B. (2009). Objective hermeneutic: Methodological reflections on social structures in women's lives. *Psychology and Society, 2*(1), 50–54.

Sullivan, C., & Cottone, R. R. (2006). Culturally based couple therapy and intercultural relationships: A review of the literature. *The Family Journal: Counseling and Therapy for couples and families, 14*(3), 221–225.

Titschler, S., Meyer, M., Wodak, R., & Vetter, E. (2000). *Methods of text and discourse analysis.* London: Sage.

Wagner, S. M., Lukassen, P., & Mahlendorf, M. (2010). Misused and missed use—Grounded theory and objective hermeneutics as methods for research in industrial marketing. *Industrial Marketing Management, 39,* 5–15.

Watzlawick, P., Beavin Bavelas, J., & Jackson, D. (1967). *Pragmatics of human communication— A study of interactional patterns, pathologies and paradoxes.* New York: Norton.

Weber, M. (1930). *The Protestant ethic and the spirit of capitalism.* New York: Scribner's.

Wernet, A. (2006). *Einführung in die Interpretationstechnik der Objektiven Hermeneutik.* Opladen: Leske+Budrich.

Chapter 9
Family Semantic Polarities and Positionings: A Semantic Analysis

Valeria Ugazio and Lisa Fellin

Introduction

How can we see the forest without losing sight of the trees? How can we see the pattern that connects members of the family without forgetting the subjectivity of each person? This dilemma has been a cause of concern throughout the history of family therapy, from its beginning until today. As Minuchin, Nichols, and Lee (2007) pointed out:

> Unfortunately in the process of stepping back to see the system, family therapists sometimes lost sight of the individual human beings that make up a family. Although it isn't possible to understand people without taking their social context into account, notably the family, it was misleading to limit our focus to the surface of interactions—to social behaviour divorced from inner experience (Minuchin et al., 2007, p. 1).

Family therapists, in truth, preferred at first to look at the system in its entirety. In attempting to identify its characteristics, the processes for constructing meaning were neglected to the advantage of pragmatic redundancies (Watzlawick, Beavin, & Jackson, 1967). The pragmatic obscured the semantic. As a result, clinical concepts and research instruments were only focused on manifest behavior. Once the mind had been equated with a black box and analysis was limited to the here and now, attention turned to symptomatic behavior, to its pragmatic effects, and to the "pol-

Translated by Richard Dixon.

V. Ugazio (✉)
Scientific Director of European Institute of Systemic-relational Therapies,
Via Ciro Menotti 11/D 20129,, MILANO

Professor of Clinical Psychology, University of Bergamo,
Piazzale Sant, Agostino 24129, BG, Italy
e-mail: valeria.ugazio@unibg.it

L. Fellin
University of East London, London, UK

© Springer International Publishing Switzerland 2016
M. Borcsa, P. Rober (eds.), *Research Perspectives in Couple Therapy*,
DOI 10.1007/978-3-319-23306-2_9

icy" the family organizes around the symptom. The aim was to describe and explain the family with concepts that transcend the individual, such as "rules," "homeostasis," "myth," "family paradigm," "structures," "boundaries," and "enmeshment" (Anderson & Bagarozzi, 1988; Ferreira, 1963; Minuchin, Rosman, & Baker, 1978; Reiss, 1981). Although useful, these concepts have a holistic nature and cannot therefore differentiate the contribution of each person in the construction of a shared family dynamic.

By the end of the 1970s, some family therapists were already distancing themselves from the model of the black box, with its exclusive focus on manifest behavior, and on the here and now. The subsequent development of constructivism and constructionism, second-order cybernetics, and, more recently, collaborative approaches have restored emotions, beliefs, and subjective experience to a respectable position. New levels of analysis have been introduced such as self-reflexivity. There has also been much talk about meaning. Nonetheless, the focus has been on family stories — understood once again as an undifferentiated whole — or on the construction of new stories in therapy. Little investigation has been made of the processes through which families, together with their therapists, construct new stories or of the quality of the new stories. Many well-known family therapists (e.g., Sluzki, 1992; White & Epston, 1990) have ended up supporting the idea that the new stories developed during the therapy were better (i.e., more helpful) simply because different from those dominant in the family. The outcome is a substantial shortage of concepts, instruments, and research paradigms that can shed light on the ways in which people create meaning together.

The method we will use to analyze the sessions with Victoria and Alfonso is focused on the semantic and inspired by the model of personality and psychopathology set out by Ugazio (1998, 2012, 2013) which reverses this trend, offering a new way of looking at the construction of bonds and problems in which they so often become entangled.

Construction of Meaning Within a Couple Relationship

Rather than focusing only on pragmatic and observable behavior, Ugazio's model focuses on semantic and analyses the processes through which each partner contributes to the construction of meaning in the couple and in the family. It offers an intersubjective approach to personality based on a constructionist conception of meaning focused on the concept of "semantic polarities," capable of accounting for the differences and similarities within families.

For the semantic polarity model (Ugazio, 1998, 2012, 2013), it is the polar structure of meaning, which seems to characterize all languages, that assures intersubjectivity, making each of us interdependent. In order to be positioned as generous, independent, cheerful, or in whatever other way, other people in the same conversational context must describe themselves and be described as selfish, dependent, and gloomy. Even beauty and physical strength — aspects whose genetic component is prevalent especially among younger people — open up opposite polar positions as soon as they become salient meanings through which the conversation develops. To feel or to be considered "good looking" or "strong" requires a certain commitment, but even to be ugly or weak requires some effort—the wrong hairstyle, the choice of clothing to enhance their own defects, or the sacrifice of any pleasant physical activity. No one makes so much effort unless beauty and physical strength are salient in their conversational contexts. Genetic attributes can of course be so devastating as to impose new meanings. When someone is born a Venus or a Hercules, it is difficult for members of the family to ignore her beauty or his strength. As soon as beauty and strength come into the conversation, some other member of the family, looking into the mirror, will discover him/herself to be an ugly duckling, while others will realize they are sickly: Hercules will have to protect them.

From this point of view, the duality is not in the individual subjectivities. Mr Hyde does not necessarily have to be inside Dr Jekyll, as claimed by Jung (1921), Kelly (1955), and more recently Hermans and colleagues (Hermans & Di Maggio, 2004). It is in the conversational context of the irreproachable Dr Jekyll that we have to search for the diabolical Mr Hyde. Meaning, according to Ugazio (1998, 2012, 2013) and Procter (1981, 1996) as well, is a joint enterprise between people who occupy a multiplicity of different positions in one and the same conversational context. And multiplicity and diversity are features of the semantic polarities. Different cultures give central importance to different semantic polarities and construct different positions inside the polarities.

One of Ugazio's central theses is that people with eating, phobic, obsessive–compulsive disorders, and depression will have grown up in families where certain specific meanings predominate.[1] For example, in a family where one member has a phobic disorder, conversation will be characterized by what is dubbed a "semantic of freedom," a dynamic driven by the emotional polarity fear/courage. Since the most relevant semantic polarities to members of such a family are freedom/dependence, or again exploration/attachment, core conversations in the family will tend to focus on episodes that centralize these polarities. As a result of these conversational processes, members of these families will feel and define themselves as fearful and cautious or, alternatively, courageous, even reckless. In families where a member develops an eating, obsessive–compulsive or mood disorder, conversations will revolve around quite different sets of meanings which Ugazio (2013) calls, respectively, the semantic of "power," "goodness," and "belonging."

The semantics referred to are not conditions sufficient for the development of the related psychopathologies. In many families where the prevailing semantic is, for example, "goodness," no one shows any sign of obsessive–compulsive disorder, even if various members of the family develop personal narratives, ways of relating and values similar to those who develop an obsessive disorder. "A crucial role in the transition from 'normality' to psychopathology is played—according to the model of semantic polarities—by the particular positions mutually assumed within the

---

[1] The hypothesis of a close relationship between meaning and psychopathology was put forward by Guidano and Liotti (Guidano, 1987, 1991; Guidano & Liotti, 1983) and later developed by other cognitivists (Neimeyer & Raskin, 2000; Villegas, 1995). All these authors, not unlike Kelly (1955), consider meaning to be something essentially individual rather than a joint undertaking.

critical semantic by the subject and by those family members who are significant to him or her" (Ugazio, 2013, p. 9).

From a semantic point of view, forming a couple signifies renegotiating personal meanings with the partner. The couple's life starts together by the meeting of two worlds of different meanings, the result of previous co-positionings. Each partner during the course of their personal story has developed particular ways of feeling and of building relationships in interconnection with the members of their own family. These are semantic patterns nourished by specific emotions.

Falling in love and the forming of a partnership are a challenge to these well-established semantic patterns at the root of our identity, but also a great opportunity for widening our meanings. At whatever stage in the life cycle these unsettling events occur, they will lead to an inevitable restructuring of meanings through emotional destabilizing moments similar to the enigmatic episodes described by Ugazio (1998, 2012, 2013, see pp. 60–66). In these episodes one partner, and sometimes both, are abruptly put inside a semantic polarity, alien to their previous "co-positionings" or are pressed to take a position which is "forbidden" in their story.

All of us are able to understand meanings present in our cultural context, even if they are different to those that dominate the conversation in our own family and in the other groups to which we belong. But our understanding of it is abstract. When we position ourselves within meanings that we only know intellectually, we are unable to interact fully because the repertory of the story we have experienced lacks adequate relational movements, forms of understanding, and ways of feeling. This is what happens in what Ugazio (1998, 2012, 2013) calls enigmatic episodes. If these episodes occur within strongly absorbing relationships, they can trigger genuine dilemmas or give rise to disruptions that can affect the relationship and even the self. Enigmatic episodes are, in fact, situations which create emotions in the partners rendering them unable to "co-position" themselves. Consequently, the future of the relationship is threatened. Although sources of discomfort and anxiety in one or both partners, these episodes are an extraordinary opportunity for learning new semantic games. The partners' emotional involvement is the best tool to overcome the impasse:

> Both partners, in attempting to continue their relationship, experience emotional state that allow them to "co-position" themselves on the basis of a semantic polarity that was previously unknown to both of them, or, as more often happens, one of the partners develops ways of feeling and relating that allow her/him to "co-position" themselves with the other (Ugazio, 2013, p. 64).

Sooner or later almost all couples will find themselves facing these destabilizing episodes.

This is particularly true when people develop their first stable relationship as a couple, as in the case of Victoria and Alfonso. It is the first inescapable confrontation with a world of meanings that is different from their own: each partner finds his or herself having to relativize their own cognitive but also emotional world. For anyone who has never experienced conflict or trauma in their own family, their semantic universe is now drastically unsettled for the first time.

A fundamental step in this process is the often traumatic meeting with the partner's family. If the love develops into a partnership, the relationship can no longer be limited to the two partners. The formation of a stable couple brings into play at least two families of origin; each partner has to renegotiate their relations with their own original family, their relatives, and their friends and at the same time find their own, more or less comfortable, positioning within the partner's family. These processes, always accompanied by episodes that are enigmatic and sometimes dramatic, can lead to the relationship ending or, conversely, to the construction of a new semantic world, which must be at least partly shared. In some cases, the semantic of one partner become the dominant universe for the couple: ways of feeling, ways of behaving, and emotions felt by a partner gradually acquire greater relevance for the other, whose original meanings recede progressively into the background. More often the two semantic worlds find some way of coming together. They become an asset for both partners: the world of each is thus enriched with new emotions and semantic games or, in the most fortunate cases, they fuse together, bringing about a new conversation in which it is hard to trace the original meanings from which it has emerged. In other cases, the couple is unable to construct a shared semantic plot: each partner remains in his/her semantic world. As a result, the bond breaks or becomes entangled in semantic traps based on misunderstandings that are primarily emotional.

Also the couple we analyze is entangled in some enigmatic episodes. The story of Victoria and Alfonso is intriguing, if for no other reason than for their tenacity in keeping their relationship going despite the many difficulties they face in their life together and the radical differences in their life plans, interests, and sensibilities. Three years have passed since, at a very young age—he was 19, she 23—they began the cohabitation that immediately brought about Victoria's depression. As well as the depression, Victoria and Alfonso have had to face language and cultural differences, and the fierce opposition of Alfonso's family, who have no intention of "losing" a son—for Mediterranean cultures still in need of family guidance—in the Scandinavian ice. What has kept these good-looking students so firmly together in an age and a culture where the possibility of dropping one relationship to start another would be quite easy? And why does not their relationship help Victoria overcome her depression? Its context is perhaps to be found in Victoria's family of origin (about which we know very little), but it persists day after day with Alfonso.

The semantic analysis that we present tries to shed light on Victoria and Alfonso's relationship, and the related puzzle, by answering these research questions.

1. With what semantics do Victoria and Alfonso interpret their own stories as individuals and as a couple?
2. Does each partner share at least part of the semantic world of the other?
3. Are the semantic games of the couple modified during the course of the four sessions? And how?

For reasons of space we will have to refrain from analyzing the therapist's contributions, which are, in our view, crucial for the therapeutic change (Ugazio & Castelli, 2015).

The Family Semantics Grid: The Analysis of Narrated and Interactive Polarities

The semantic analysis of the four sessions with Victoria and Alfonso was carried out by two independent coders, applying the Family Semantics Grids (Ugazio & Guarnieri, in press; Ugazio, Negri, Fellin, & Di Pasquale, 2009). Inspired by the semantic polarities model, the FSG distinguishes and identifies two kinds of semantic polarities: narrated and interactive. The first are the semantic oppositions within which each partner, during conversation, defines him or herself, other people, and events. These polarities construct the "narrated story," which may be different from the "lived experience" that each conversational partner puts into action when interacting with his or her interlocutors. For example, a client, while describing himself as a victim of his wife's harassment, may use an assertive tone that places him in the position of a prosecutor, committed to winning the support of the therapist so as to get his wife into the dock. The interactive polarities are discursive phenomena of a performative type. They are inferred by the way in which the two partners and their therapist reciprocally position themselves, often implicitly as their conversation unfolds. Of course, therapists also take position and are positioned by their patients, often ending up co-positioning themselves in the semantic that dominates the family conversation. In a family session the therapist can, for example, position herself as a secure base that allows all members of the family to discuss a conflicting argument, whereas just after she can seek to make the voice of one member of the family heard, offering herself as that member's (temporary) ally.

The FSG, as well as identifying the polarities, enables them to be classified into the four semantics mentioned in the previous section (p. 3). Each semantic is operationalized by a grid of 36 polarities, fuelled by the same emotional opposition. The classification also makes it possible to distinguish the area of social construction of reality (emotions and sensations, ways of relating, definitions of self/others/relationships, values) to which each polarity refers. Figures 9.1 and 9.2 summarize the two grids we will be most using to examine the sessions with Victoria and Alfonso: those of the semantics of "freedom" and "belonging."

The FSG is a qualitative system with inferential aspects that distinguish it from computer aided methods (CAQDAS). For the coding of narrated polarities, inference is limited to three steps:

(a) Identification of the second pole of each polarity (whose exploration, in the case examined here, extended to four sessions);
(b) Identification of the "operative definitions" of each pole, namely, the expressions that provide the most concrete definition of each pole inside the verbatim of the four sessions;
(c) Reframing of the semantic content of each pole in the light of the opposite pole, of the "operative definition" of the poles and on the basis of conventional meanings. This eliminates misunderstandings arising from the use of idiosyncratic expressions, which are frequent in therapy conversation as in all informal conversation. Only after this redefinition does the method make it possible to pass on to the classification of each polarity in accordance with the grids.

VALUES	
FREEDOM	DEPENDENCY
EXPLORATION	ATTACHMENT
RISK	SAFETY
CHANGE	STABILITY
DEFINITIONS OF SELF/OTHERS/RELATIONSHIPS	
FREE	DEPENDENT
Self-sufficient	Conditioned by others
Explorative	Trapped
Unbond	Commited
UP AGAINST THE ODDS	PROTECTED
Nomadic	Sedentary
Precarious	Stable
Disoriented	Safeguarded
UNPREDICTABLE	RELIABLE
Distant	Close
Stranger	Familiar
Dangerous	Reassuring
COURAGEOUS	FEARFUL
Rash	Cautious
Careless	Careful
Bold	Cowardly
STRONG	WEAK
Invulnerable	Fragile
WAYS OF RELATING	
KEEPING DISTANT	GETTING CLOSE
Counting on oneself	Counting on others
Opening to others	Closing others out
GETTING FREE FROM OTHERS	DEPENDING ON OTHERS
Breaking free	Clinging to others
Keeping self-sufficient	Relying on others
EXPLORING	STAYNG PUT
Opening to novelty	Digging in
Taking risks	Protecting oneself
SCARING	REASSURING
Disorienting	Guiding
Alarming	Calming
EMOTIONS AND FEELINGS	
COURAGE	FEAR
DISORIENTATION	CONSTRAINT

Fig. 9.1   Semantic of freedom grid (from Ugazio et al., 2009)

Inference is certainly greater in detecting the interactive polarities than in identifying the narrated polarities. The former are mostly nonverbal and therefore should be identified watching the video-recording of the sessions, whereas the latter should be extracted from the verbatim transcripts. The coding system developed by Ugazio et al. (in press) first of all divides each session into relational configurations lasting between 2 and 10 min. The configurations, expressed in graph form, show the type and intensity of partner involvement, agency, and emotional climate following a method taken partly from Hinde and Herrmann (1977). Two types of interactive semantic polarities—"macro" and "micro"—are identified within each configuration. The first corresponds with relatively stable positionings (at least 2′) between patients and therapists and/or between the patients themselves within one and the same dimension of meaning. The micro-interactive polarities relate to shorter positionings (less than 2′) that stray from the meaning of the macro-interactive

VALUES		
INCLUSION		EXCLUSION
HONOR	⇄	DISGRACE
BEING CHOSEN		BEING REJECTED
GLORY	⇄	DOWNFALL

DEFINITIONS OF SELF/OTHERS/RELATIONSHIPS		
IN THE GROUP		OUT OF THE GROUP
Belonging		Excluded
Being welcomed		Being discarded
Accepted		Kept out
WORTHY	⇄	UNWORTHY
Respectable		Contemptible
Honorable		Despicable
Deserving		Reprehensible
ELECTED		OUTCAST
Rewarded		Deprived
Respected		Refused
Revered		Defrauded
GRATEFUL		ANGRY
Enthusiastic		Miserable
Joyful		Inconsolable
Merry		Hopeless
ENERGETIC	→	RUN DOWN
Together (with)		Alone

WAYS OF RELATING		
INCLUDING		OSTRACIZING
Sharing		Cutting off
Welcoming		Abandoning
HONORING	⇄	DISHONORING
Deserving		Usurping
Ennobling		Discrediting
OVERWHELMING WITH GOODS		DEFRAUDING
Remembering		Forgetting
Celebrating		Ignoring
VENERATING		DESTROYING
Jubilating		Getting down
Repairing		Regretting

EMOTIONS AND FEELINGS	
JOY	DESPERATION/ANGER
GRATEFULNESS	RESENTMENT

Fig. 9.2 Semantic of belonging grid (from Ugazio et al., 2009)

polarity within which they develop without modifying the relational and emotional climate, or re-establishing it immediately after having modified it. They can also function as a transition to a new macro-positioning. Once they have been described in detail, the micro- and macro-interactive polarities are classified according to the grids.

The coding procedure is detailed in the Family Semantics Grids (Ugazio et al., 2009, in press). Reliability, tested on 20 % of the transcripts and videorecorded four sessions, by Cohen's K between independent coders is 0.75 for narrated polarities and 0.66 for interactive polarities.

The Analysis of the Couple Sessions: Two Semantic Worlds in Search of a Co-position

A Conversation Between Freedom and Belonging

The semantics of belonging and freedom dominate the four sessions (Table 9.1) and it is the former that prevails (38 % against 23 %).

The main author of the narrated story is Victoria, who introduces most of the narrated semantic polarities (60 % against 24 %). Much of her dominance is nevertheless "forced"[2]: it is Alfonso who, in attempting to withdraw from semantic involvement and from defining the underlying relationship, induces her to provide the semantic framework. Alfonso also introduces meanings into the conversation, but mostly in response to explicit or implicit questions from Victoria and the therapists. Even when he and his family are discussed, he rarely takes the initiative in introducing new meanings that correct, contradict or further explore those introduced by his partner. Indeed, his answers are sometimes so mumbled and fragmented that it was difficult to transcribe, understand, and code them. But Victoria does not introduce both dominant semantics into the narrated story. The semantic of freedom is introduced mainly by Alfonso (48 % of his narrated polarities) and that of belonging by Victoria (50 % of her narrated polarities).[3]

These quantitative results enable us to draw an initial conclusion: Victoria and Alfonso are a couple formed on different semantics. But what specific meanings, among those characteristic of these two semantics, organize their conversation? Victoria contributes toward constructing the story of the couple and their relationship with Alfonso's family using essentially four polarities: involvement in an all-absorbing love/feeling rejected; sharing/being ignored or misunderstood; belonging/feeling excluded; being at the center of her partner's emotional world/being forgotten, abandoned. For Victoria, their relationship ought to have first place and each ought to be the center of the other's life, even when they are physically apart (I, 135;

Table 9.1 Narrated and interactive (macro + micro) semantic polarities during the four sessions

Semantic	Narrated SP					Interactive SP				
	V	A	T1 + T2	Total	%	V	A	T1 + T2	Total	%
Freedom	34	50	16	100	23.4	4	21	2	27	13.2
Goodness	5	0	0	5	1.2	0	0	2	2	1.0
Power	6	2	0	8	1.9	1	0	0	1	0.5
Belonging	129	16	19	164	38.4	30	4	4	38	18.6
Others	83	36	31	150	35.1	50	35	51	136	65.8
TOTAL	257	104	66	427	100	85	60	59	204	100

---

[2] As emerges from the analysis of the video recordings.

[3] The polarities of the other two semantics (goodness and power) are very few indeed (Table 9.1).

II, 77). This all-inclusive relationship ought to give Victoria a guaranteed feeling of belonging. Victoria talks several times about her and Alfonso as a "family" and she as his "girlfriend," emphasizing the obligations this position involves for the partner (I, 139, 219; II, 366, 376). "I need," says Victoria, "to feel that I am your girlfriend and I am important to you, that you care, that's the point" (I, 219). Yet Victoria feels she is of last importance to Alfonso, who is not prepared to share even the "small things"—this is what she calls the small misunderstandings in their relationship— who does not help in running the house, who does not invest time and money in their life together. She repeats obsessively "he doesn't love me anymore" (I, 228 or IV, 83), indeed "he hates me" (I, 228 or IV, 83 and 93), "I feel like rejected" (I, 188), "I just feel, just I am nothing for him" (II, 77). As well as not feeling she is Alfonso's girlfriend (I, 145, 174), Victoria experiences painful exclusion from his family of origin. She cannot co-position herself among the "Bold and Beautiful," as she has nicknamed Alfonso's family. Though she has tried to be included in this family, visiting them at least six times, she is now convinced it is an impossible objective: they are ontologically different. She has, therefore, severed all relations with them.

Consistent with these meanings are the three polarities most frequently introduced by Victoria to define herself: "alone/together"; "rejected/welcome"; "unworthy/respectable." Victoria repeats obsessively that she does not feel "together" with Alfonso and feels rejected—for her, as she explains to the therapist, "rejected means abandoned" (I, 194). The unworthy/respectable polarity is introduced almost exclusively in narrating her relations with the "Bold and Beautiful." Being with them, for Victoria, means losing all worth (I, 318, 331). Her self-esteem collapses when she meets them. This is a point we will return to later. Unlike Alfonso, Victoria talks much about what she feels and can describe her emotions in detail. It is true that she uses the vague expression "to feel bad/to feel good," but she fills it with meanings denoting sadness, desperation, and anger, emotional states establishing the semantic of belonging, typical of mood disorders (Ugazio, 2012, 2013). Also the interactive polarities acted out during the sessions, as we shall see, are equally characteristic of the semantic of belonging.

Keeping distant/getting close, being detached/committed, being autonomous/ needy, being strong/afraid, and being patient/not caring are the polarities most frequently emerging from Alfonso's narrated story. There are also many interactive polarities that indicate his movements away from and closer to his partner during the session. These movements, at the level of narrated story, more often concern him, his family of origin, and friends. As we have already said, Alfonso tends to be semantically cautious in his relationship with Victoria. "Being patient, listening/ being exhausted, tired," his most repeated narrated semantic polarity aimed at the couple relationship, underpins the negative pole: "I don't care," explicitly used toward other targets (his family of origin, the couple's house). Alfonso also defines himself within two polarities of the semantic of freedom: "free/controlled," or even "trapped," and "distant/attached."

Alfonso describes his family as "strong and very close" (I, 328), but is careful to emphasize that he is not too closely tied to them, to have been on poor terms with them before leaving home (I, 345). He and his family get on better now since he is in the position of a "tourist," as he jokingly puts it, when he goes back to visit them. His narrated story sees him committed to defending his "freedom" and "independence" (I, 345; II, 273). Being free is the most important thing for him: all the rest falls into an area of indifference, expressed by a persistently repeated "I don't care."

An Example: The Semantic Exchange That Begins the Therapy

The two semantic universes of Victoria and Alfonso already stand out at the beginning of the first session. Let us look at some of the key moments.

Victoria starts the conversation describing herself as depressed. Here her depression is presented as an external event that has negatively affected the couple relationship, leaving wounds still open (I, 10). Starting from this passage—which introduces the "wound/repair" polarity, with the therapist in the position of someone who has to repair the relationship (as confirmed at I, 390)—Victoria makes an interesting reversal: "…and I think also the reason why in that point of my life I got sick was, was because it was so difficult to start trusting someone, to feel loved and to feel love" (I, 10). The depression is no longer an external occurrence, nor is it even responsible for the deterioration of the couple's relationship. On the contrary, Victoria seems to have become ill because it is difficult "to start trusting someone," and above all "to feel loved and to feel love." By introducing this last pole, Victoria opens the doors of her emotional world: for her it is essential to feel loved. But how is this possible if Alfonso is not even prepared to discuss their little misunderstandings (I, 16–17)?

Alfonso also takes us immediately into his semantic with the polarity "being strong/reacting too strongly," which is soon transformed into "being strong/being frightened." He constructs a past in which he was "strong" as well as "patient" in listening to Victoria, which contrasts to a present in which he reacts "too strongly." When the therapist tries to get him to define what he means by "reacting too strongly," Alfonso hesitates. There is a long pause which ends with Alfonso whispering, almost stammering, "irritated," immediately corrected by Victoria, who redefines his emotional state as fear—"he is afraid"—and a few rounds later as "panic" (I, 22–28).

This "he is afraid," with which the central emotion of the semantic of freedom bursts into the session, would seem an abuse on the part of Victoria. If we look only at the transcript, Victoria would seem to be taking over from her partner in defining his emotions. In reality, when the episode is examined in the context of the nonverbal behavior characterizing it, a micro-interactive polarity appears in which Alfonso is relying on her to assume the responsibility of defining his state of mind. Alfonso stammers and stumbles, clearly in difficulty of answering the therapist's questions. Even when, after much hesitation, he utters that word "irr…itated," he keeps requesting Victoria's help with his eyes. By intervening, she seems to free her companion from his stress, and he seems relieved. Alfonso seems paralyzed with fear from the beginning of the session. The new situation he faces seems to make him very anxious: Alfonso does not look at the therapist for almost a minute, looking downward. When the therapist unequivocally opens up the encounter, he looks up but hides his hands further into the sleeves of his sweater as children do, places them in front of his genitals and leans back against the chair. He remains fixed rigidly in this posture until turn 34, which closes the first macro-interactive polarity.

Difficult Positionings

What position do Alfonso and Victoria assume within the two semantics that dominated the conversation?

In her narrated story, Victoria occupies the negative pole of the semantic of belonging, while Alfonso and his family are placed at the positive end. She feels "rejected," "abandoned," even "hated." Alfonso, on the other hand, is loved not just by her but also by his family of origin. "The Bold and Beautiful" love him and they love each other—Victoria acknowledges this, though begrudgingly (I, 329; II, 398). The lived story during the sessions confirms these positionings but gives greater agency to Victoria. Victoria is not in the passive position of one who feels rejected and waits to be included by others. On the contrary, she presses her partner with direct and indirect demands for a loving commitment and exclusive involvement that Alfonso regularly fails to give. Even in the here and now of the sessions, Victoria seems to be looking for a sharing with her interlocutors. The therapist offers it to her, sometimes through brief exchanges in their native language, Alfonso does not.

The most difficult positioning for Victoria, where her account becomes more somber and dramatic, is within the "unworthy/respectable" polarity, introduced by Victoria almost exclusively in relation to Alfonso's family. The "Bold and Beautiful," and above all Alfonso's mother, make her feel a horrible person, worthless, "this trash" (I, 331). Victoria feels hated by this woman, described as a Mediterranean mother who "wants to own her family" (I, 340), "because I stole her son" (I, 342). It is an important passage. The rejection felt by Victoria is not one where a mother dislikes her son's girlfriend. She feels accused of stealing the son. But how could Alfonso's mother make her feel like a thief unless she regarded her son as a child? Being equated with a child-snatcher makes Victoria feel unworthy and irritates Alfonso. It puts him back in that position of a young boy who needs the guidance and protection of his mother from whom he escaped when he went to Scandinavia on an Erasmus Scholarship and extended his stay there by 3 years. Alfonso in fact intervenes, emphasizing that his mother and his relatives "think" they possess him, whereas he has broken away from them (II, 468).

It is possible that the cause triggering off Victoria's depression—which she dates back to her first visit to Alfonso's family—was this very perception of herself, as unworthy, given to her by Alfonso's mother. Indeed she left the visit crying desperately during the whole journey back (II, 320). It is a perception that risks creating an intransitivity between her honor and her relationship with Alfonso and throws her

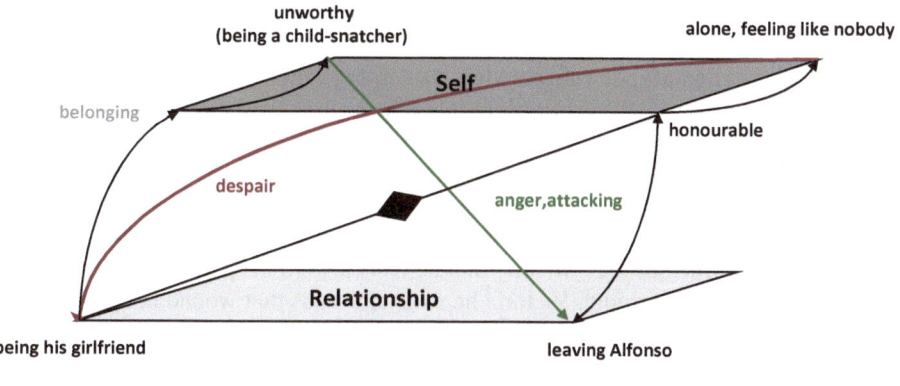

Fig. 9.3 Victoria's dilemma

into a dilemma (Fig. 9.3), similar to a 'strange loop' (Cronen et al., 1982). Continuing her relationship with Alfonso gives her a sense of belonging: although she feels inadequately treated by him, she is still his girlfriend. But this belonging makes her feel like a thief and so contemptible. On the other hand, leaving Alfonso, or causing a rift in the relationship by increasing the level of conflict, would consign her to despair. Being alone, for Victoria, is the same as feeling nobody, as she declares several times. And Alfonso seems to be all she has. From a bitter comment (I, 367) we know that Victoria's family of origin is absent from their life, though we do not know why—her family story is one of the main unspoken issues (Rober, 2002) in the sessions. Nor is there any mention of Victoria having friends and Alfonso's friends do not seem to be shared jointly. Alfonso's adolescent behavior contributes to the dilemma. With his desire always to be out with his friends, with his scarce involvement in domestic matters and disinterest in the house, Alfonso seems to treat his companion more as a mother than as his girlfriend.

In this more interpretative part of our analysis, we have redefined "steal" as "usurp" due to the presence of a strong difference in their respective plans at this stage in their life as well. He is in his 21, she is 4 years older. Victoria talks about them as a "family" and about the possibility of getting married and having children. Alfonso uses the word family only for their respective families of origin, and seems much more interested in friends and exploring the world than in building a family. He also comes from a Mediterranean country where it is unusual, especially for a male, to start cohabiting at the age of 19, as he did. Young people who study, including women, start living with a partner around the age of 30 as a rule.

But let us look at Alfonso's positioning. He describes himself as free and independent, able to avoid being influenced by his family, from whom he is very happy to be far away, determined to defend his own space and substantially uncommitted in his relationship with Victoria. They are positionings all around the positive pole of the semantic of freedom. On the other hand many of the interactive polarities show him to be frightened and afraid. This position also emerges in his narrated story when he speaks of his reactions to his partner's requests for love, consolation, commitment. He presents them as symptoms and judges them negatively as "excessive," he repeats several times "I have such a reaction," "I can't deal." With the help of the therapist and of Victoria (I, 20–29) it emerges that he is frightened. These emotions that make him feel "fragile" and "weak" are probably at the root of the "excessive reactions" that worry Alfonso. Much of the difficulty that he feels toward his partner could be traced back to this situation: through Victoria he feels emotions that wound his self-esteem, since they put him back into that position of fragility that he probably felt in his family and from which he broke free by moving 2000 km away.

Meanings Misunderstood: "To Be Between Two Families"

Victoria and Alfonso frequently misunderstand each other because each uses their own semantic to interpret the same sequence of events, or even the same positioning. Emblematic is the misunderstanding concerning Alfonso's positioning that Victoria describes as being "between two families." We are in the first session (I, 364–365). The conversation focuses on Victoria's break with Alfonso's family, whom Alfonso will soon be going to visit alone for a week. Victoria concentrates on the aspect of the problem closest to her heart—Alfonso's loyalties after the rift—and puts forward the idea that Alfonso is in "the middle of two families: me and his family." The metaphor, already introduced a little earlier (I, 139), implicitly institutionalizes their relationship, and moreover in equal terms to his family of origin: they too are a "family." This is an un-negotiated positioning: Alfonso never describes the two of them as a family. Nor, when he talks about them, does he ever use the expression couple or describe Victoria as his girlfriend. Alfonso avoids all words that allude to some form of institutionalization of their relationship, which are abundant in Victoria's account. Victoria, in reintroducing the metaphor, adds all her regret that she has put her companion into such a hard position: "I know that for Alfonso's home-country family is very important and I hate that he is in between two families, because it's very tough for him" (I, 364). Victoria feels she must show she understands Alfonso's supposed pain, that she is sorry she cannot get on with his family, and above all she wants to throw responsibility for the rift onto the "Bold and Beautiful." It is a rift that makes not only her suffer (which is of little importance to them) but also their son. For this, she accuses them of being selfish.

Victoria's account aims to make Alfonso understand that the best "family," the one he must choose, is *her*. But this is not just theater: Victoria is really convinced that this rift is hurting Alfonso so much that it makes him want to leave her. In her semantic, the conflicts of loyalty are devastating: they threaten those sought-after bonds on which your worth depends. This being "in between" is, on the contrary, a positive solution for Alfonso: it gives him the independence he so yearns. Sure of his family ties, which he has not the slightest intention of putting at risk,[4] this solution enables him to keep his parents at a distance, ensuring that they do not come to visit him. In Scandinavia, free from their control, he can therefore develop his own independence, which was under threat while he lived with them. By going to see them it is also easier to maintain a good relationship with them and distance himself, even temporarily, from Victoria. Alfonso's problem, which is typical of those dominated by the semantic of freedom, is in regulating the distances and keeping control over this process (Ugazio, 1998, 2013). It is not therefore just to comfort Victoria that he declares he is not suffering and sees certain advantages in the rift ("now I like to be, to have my freedom somehow, so if they don't come here, it's really fine by me," I,365). Being in between, even if it is not the ideal position, is the best position for him at this stage in his life.

Are Victoria and Alfonso therefore incapable of entering into the semantic of the other, as this episode would seem to suggest? Our analysis does not allow this conclusion. Victoria is able to enter at least partly into Alfonso's semantic. When the target is her partner and his family, Victoria uses the semantic of freedom almost as much as her own dominant semantic (22 vs. 27). There is, for example, a sequence (I, 344–352) in which Victoria, moved by the desire to detach her partner from his Mediterranean family, seems even to talk with his voice. She gives him a description of the emotional atmosphere of "constraint" and "control" in his family, which is abhorrent to anyone like Alfonso who is positioned within the semantic of freedom. For people in this position "being possessed," "put under pressure," "controlled" produce anxiety, loss of control, and the risk of being swept away by emotions and placed in a position of weakness.

Victoria can also "close"[5] certain narrated semantic polarities, opened by her partner, that are characteristic of the semantic of freedom. This happens, for example, in session III where she "closes" the "searching for freedom and independence" pole, introduced by Alfonso, with being "trapped" (III, 148), a meaning typical of the semantic of freedom which perfectly expresses her partner's way of feeling. On the contrary, Alfonso seems unable to tune into the same semantic wavelength as Victoria. Alfonso almost never uses the semantic of belonging, even when the target is Victoria. On the few occasions that he redefines a pole introduced by his partner, it is generally through of his own semantic. Victoria is nevertheless only partially able to enter Alfonso's semantic world: when she speaks about her relationship with her partner—the subject closest to her heart—or about herself, she rarely uses the semantic of freedom. Here it is the semantic of belonging that dominates.

---

[4] The most that Victoria manages to make him say is that he has distanced himself a little further from his family after his partner's rift (II, 468).

[5] Each polarity has three poles, the two extremes and the intermediates, that can be summarized as the middle point. For example between love and hate there is a range of intermediate sentiments that Ogden (1932) encapsulates in the middle point of indifference. Speakers introduce one pole at a time during the conversation and different actors can express the other two poles of each polarity. In this analysis, we consider a polarity "closed", when at least two poles are verbalized during the four sessions, whereas "open", when we cannot find any complementary pole in the four sessions.

Does the Lived Experience During the Sessions Confirm the Narrated Story?

Even though the semantics expressed by Victoria & Alfonso during their lived experience in the here and now of the sessions are the same of their narrated story, the specific polarities and their relative positionings are often different. Here is one example. During the sessions, Victoria introduces the polarity "attack/repair" which, though a characteristic of the semantic of belonging, does not figure among her narrated semantic polarities. It typically co-positions with Alfonso's most frequent interactive polarity: "keeping distant/getting close." Their positionings inside these two polarities create a characteristic pattern that emerges for the first time at the beginning of the first session (turns 146–178) and is repeated several times in the next sessions.

The pattern illustrated in Fig. 9.4 can be described as follows. Victoria asks for greater commitment from her partner, who interprets her request as an attack. He, therefore, moves physically away from her, moving his body to the other side of the chair, he stops looking at her, and becomes tense. Faced with this feedback—in Victoria's semantic a rejection—she softens her tone, often through micro-interactive polarities that reduce the tension and act, in this context, as a repair. Experienced by Victoria as an extreme and humiliating attempt to save their relationship, these acts of repair come as a relief to Alfonso: they reduce his partner's pressure over him. Relieved, Alfonso moves back toward Victoria, trying once again to catch her eye and smiling at her. Unfortunately, the reconciliation is short-lived. Alfonso's move back toward her—which she judges to be "forced," brought about by her (degrading) initiative—is immediately followed by another request/attack by Victoria. She wants to

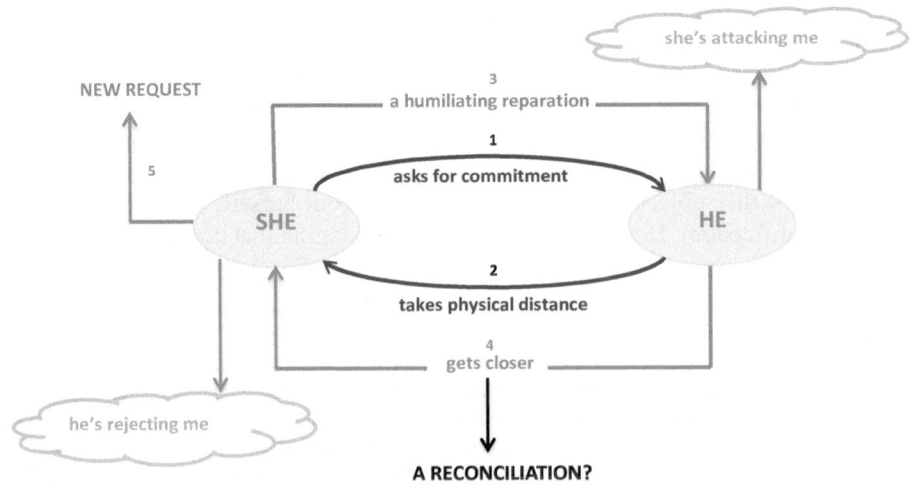

Fig. 9.4 One of the couple's recurrent semantic pattern

make sure he loves her or, at worst, to recover some of her self-respect, lost by forcing her partner into reconciliation with her. Alfonso, who has just moved back toward her and expects some peace ("I need a rest," I, 74), feels pursued once again by a new request/attack from the partner and cycle starts again.

What does this pattern reveal? One key aspect: Victoria and Alfonso manage to complement each other at an interactive level and perhaps this is why they have not left each other, despite the difficulties in their relationship, even though both are at an age when other relationships are easily possible. But there is no *semanticcohesion* between them. In other words, they develop a sort of semantic dance that brings exhaustion and frustration on both sides since each is moving to a different step, in time to different music. Their interaction does not seem to create a shared semantic score.

Do the Couple's Semantic Games Change During the Consultation?

To answer this question, we shall look at the semantically most interesting part of the fourth and last session (IV, 180–190, 206–207): the discussion on the value that each gives to the place where they live, a theme made relevant by the imminent move of house.

Discussion on this theme immediately becomes a metaphor for their relationship which brings two main polarities into play: "home/house"[6] and "committed/unbound" (Fig. 9.5). For Victoria, the place where they live is a "home," meaning a "warm nest" where you feel you belong ("It doesn't feel like home when there's no carpets," IV, 180). For Alfonso, it is simply a base for exploration, so that it is a "house," meaning "any kind of hell hole" (IV, 206) close to the center and therefore convenient for going out. Committed/unbound is expressed not only by the differing importance that the partners give to the metaphorical object, the place where they live, but also the financial investment and time that are prepared to ascribe to it. Alfonso wants to spend as little as possible on furnishing the house, nor does he intend to devote any time to decorating or cleaning it, whereas Victoria wants to invest time and money in making the house as comfortable as possible.

The exchange shows that the couple still lacks semantic cohesion. The two polarities they create are derived from semantic worlds that are incapable of moving closer together and of transforming. Victoria shows also here that she is able to enter Alfonso's semantic world; it is she, for example, who defines what her partner considers to be a house: "any kind of hell hole," provided it is close to the center. Alfonso does not go into her semantic. Still there is a small and important development in his narrated story: he begins to understand her meanings and also to respect them. He

---

[6] Due to the symbolic value it assumes during the conversation, this narrated polarity is considered as related to the area of values and is redefined as "belonging/exploring".

NARRATED POLARITIES

Pole 1					Pole 2					Polarity		S.G. Cod.	
										Ridecclination			
Semantic Area	Operative Definition	Turn	Atb	Tg	Semantic Area	Operative Definition	Turn	Atb	Tg	1	2	1	2
Alfonso is very bad in decorating the house	because for him a "hell hole" in the centre is enough (206)	151	1	2	but to me it doesn't feel like home when there's no carpets	home is a warm nest	181	1	1	nomadic	belonging	122	420
I have always have to fight my ways even if they are the right ones		151	1	1	If I say, 'Ok let's, we can just use the furniture that we have' for example, but then she insists let's buy a new something new, I think at the end that's		207	2	2	fighting	resisting	334	331
it's, like,it would our turn to go there but since I am not ready he's going (alone)	she is not ready to stand the exclusion by the Bold and Beautiful so she prefers to self exclude herself from them	170	1	1						cutting off		431	
to me a house it's ok, it's not the most important, like for her it's really important	to me home it's more important if it's just easier to live like with, (if it)gives less problems (191)	185	2	2	but for me home is like the most important thing		190	1	1	exploration	belonging	112	410
we move to this new place and maybe I don't think that we should buy some new stuff, to me, it's it's; I think maybe it's just not so important to me	I wouldn't spend much money on decorating it or this kind of thing and also I would keep it maybe more simple, so nothing, more practical, so that it's more simple to may be clean it or live in it(193)	189	2	2	(...) that I want to get you in a trap (session III)		148	1	2	to be not commited	trapped	110	111
I think we shouldn't have moved now may be,that we would have found some flat later, but then in the end we talked about it and we decided to, but, yeah for example I'm more, for example, like, that maybe it's easier, to me it's better to find a place, because that is a really nice place, like in the house inside, for example you know like the wooden floor the walls, but maybe to me it's, that it may be more important may be that it's in, closer to the centre	house has to be near the centre, so I'm free to go out alone	199	2	2	that you would move to any kind of hell hole as long as it's in the centre but I can't	it's better to have a warm nest out of the city instead of a "hell hole" in the centre of the city	206	1	1	nomadic	belonging	122	420

Fig. 9.5 Narrated semantic polarities from 132 to 199, session IV. Atb, attributor; Tg, target; 1, Victoria; 2, Alfonso; 3, T1; 4, T2; S.G. Cod., semantic grid codes: 110–143 semantic of freedom, 210–243 semantic of goodness, 310–343 semantic of power, 410–443 semantic of belonging, and 510–540 other semantics

emphasizes, for example, that the home is really important for her (IV 189, 192) and likes the new house that Victoria has chosen: "it's really a nice place" (IV 198, 203).

Do any greater changes emerge in the session lived story? Even if the relational climate is less tense, the prevailing way for constructing meaning between Victoria and Alfonso remains the same: she asks for her partner's commitment (which, in their dynamic, assumes the meaning of attacks). Alfonso replies with avoidance behaviors (changing the subject, moving his posture away from Victoria, or turning to the therapists), at which Victoria becomes conciliatory, underplaying the problem or declaring that she is the one responsible for their difficulty since she is "sick." This dynamic is maintained not just when Alfonso moves away but also when he rejects her, as happens in a micro (12-s) interactive polarity (IV, 167–169). "Can you really do without me at Christmas?" asks Victoria while they are discussing the forthcoming Christmas holidays, which they will presumably be spending 2000 km apart, each in their own country. Alfonso's "yes" sounds like a rejection to Victoria, who immediately tries to re-establish the relationship with Alfonso (who is visibly annoyed), with a relational movement of repair: the problem is hers; she is not yet ready to face Alfonso's parents.

Yet, Alfonso and Victoria, for the first time in the fourth session, create a new interactive polarity that expresses a sharing between them (Fig. 9.6). During the other three sessions, Alfonso had never shared any meaning with Victoria, he had never entered her meanings: he remained outside them, both analogically as well as verbally, often frightened. But what do Victoria and Alfonso share? Their differences! Their semantic worlds remain distant. Alfonso still does not position himself within Victoria's semantic, but now it does not confuse him, he does not move away when faced with meanings introduced by Victoria. On the contrary, Alfonso shows, even analogically, that he understands the way she feels, even if it is a way of feeling that is not his, as he explains to the therapist. Victoria, for her part, shows an equally empathetic understanding of Alfonso's meanings. It is an interactive polarity of almost 2 min and almost completely covers one relational configuration in which the exchange is between the couple, while the therapists, for the most part, play the role of active observers. There are just two brief overlapping micro-interactive polarities. The first, of just a few seconds, opens the configuration and arises from Alfonso's initiative. For the first time he speaks with Victoria's voice, demonstrating that he understands her meanings and positioning. The second occurs just before the closure of the macro-interactive polarity. This one too expresses a brief sharing, but this time its protagonists are the therapist and Victoria. What are they sharing? Their identities! Faced with Alfonso's claim that the area where they are going to live is suburban, the therapist agrees with Victoria: how can it be suburban? It is only a few steps from the center! Victoria and the therapist, both Scandinavian, share an idea of distance that is very different to Alfonso's Mediterranean idea. For him, 800 m in the Scandinavian ice might be enough to obstruct him from reaching the center. The therapist immediately jokingly distances himself from this brief lapse in favor of Victoria by turning to Alfonso: "Oh did I say something wrong?" (IV, 196). Nevertheless, these few seconds of shared cultural identity between him and Victoria acquire a symbolic value precisely because they occur when the two partners become aware that all they can share are their differences.

The sessions have also led Alfonso to enter Victoria's semantic world: now they can share. But it is not the happy ending that Victoria had desired. She has, not surprisingly, been rather sulky during most of the session. As emerges from the self-reports, Victoria is satisfied with the consultation but is coming to realize that sharing, for her and Alfonso, means feeling the profound difference of their semantic worlds. The few seconds of sharing with the therapist seem to leave open to Victoria the possibility that the sharing she so yearns for might be rather easier in other relationships, with people more open to construct joint semantic worlds.

Conclusions

The analysis illustrated here highlights that Victoria and Alfonso present a particularly low "semantic cohesion" during the sessions. Their worlds are dominated by different semantics that do not come together. The small amount of semantic cohesion between them is assured asymmetrically by Victoria. It is she who enters her companion's meanings when there is discussion about him or his family. It is she who is able to "close" certain polarities by introducing meanings, characteristic of the semantic of freedom—Alfonso's semantic. And it is also she who creates interactive polarities, especially of protection and guidance, typical of his semantic. Alfonso seems unable either to enter or interact with Victoria's semantic. Nevertheless, Victoria herself does not utilize her companion's dominant semantic when the discussion is about her or their relationship. The encounter does not seem to have widened Victoria's own semantic horizon very much: when she has to express her own personal and relational experience, she too remains solidly anchored to a semantic world that belongs to past or perhaps present co-positionings with other conversational partners.

The specific semantics through which Victoria and Alfonso contribute to the construction of the conversation have also been identified. Victoria introduces the semantic of belonging, typical of depressive disorders. This is a result that further[7] confirms the link hypothesized by Ugazio (2012, 2013) between depression and the semantic of belonging. Instead, Alfonso reads events and interacts through the semantic of the freedom. The meeting of these two semantics creates many misunderstandings and dysfunctional interactive patterns, two of which we have analyzed in the previous section.

One of the aims of the applied method is precisely the identification of "semantics" and the evaluation of the "semantic cohesion." Operationalized by the method used here for the first time, these two variables fit particularly well with research on multiactor sessions. We have already widely discussed the "semantics"

---

[7] See Ugazio et al. (2015).

MACRO INTERACTIVE POLARITIES

Turn	Configuration	Analysis	Pole 1 Atb	Pole 1 Tg	Pole 1 Relational Movement	Pole 2 Atb	Pole 2 Tg	Pole 2 Relational Movement	S.G. Cod.
132–141	(diagram)	T1, after a silence, manages to involve both partners. V tries to involve A analogically (turning towards him) while she 'attacks' him by introducing the hot topic (A's parents don't aknowledge their relation). A takes distance and changes topic.	1	2	attacking in a sorrowful way	2	1+3	keeping distant	530 / 130
142–160	(diagram)	V moans grumpily, while hiding and tormenting her hands in her sleeves; A avoids the attack making his point kindly	1	2	complaining	2	1+3	self-justifying	530
161–171	(diagram)	T2, for the first and only time, intervenes, shifting the focus on the abandoned hot topic. A replies at ease, V ignores her by replying to T1 and accusing A.	4	2	including	2	4	including	430
						1	4	excluding	431
172–184 (half)	(diagram)		1	2	"assailing"	2	1	taking one's lump	530
			3	1+2	consulting	1	2	downplaying	530
184 (half) 197	(diagram)	A and V share their differences. It ends with laughter.	2	1	sharing	1	2	sharing	430

MICRO INTERACTIVE POLARITIES

Turn	Analysis	Pole 1 Atb	Pole 1 Tg	Pole 1 Relational Movement	Pole 2 Atb	Pole 2 Tg	Pole 2 Relational Movement	S.G. Cod.
153	T1 jokes about the carpets, V raises a forced smile and snuggles up into her jumper looking down	3	1	downplaying jockingly	1	3	lightening momentarily	530 / 530
167	V looks up to A, and speaking only to him, she asks "Can you really stay without me?" He analogically says 'yes'	1	2	amorous request	2	1	rejecting	530 / 530
169	V remedies by taking responsibility of the problem	1	2	repairing				436
176	T1 is ironic, V laughs.	3	1	downplaying jokingly	2	1	calming down	530 / 530
184–186	A looks at V and speak with her voice	2	1+3	being able to understand				530
193–197	V includes T1, they both share the nordic sense of physical space.	1	3	including/sharing	3	1	sharing	430 / 430

Fig. 9.6 Interactive polarities from 132 to 199, session IV. Atb, attributor; Tg, target; 1, Victoria; 2, Alfonso; 3, T1; 4, T2; S.G. Cod., semantic grid codes: 110–143 semantic of freedom, 210–243 semantic of goodness, 310–343 semantic of power, 410–443 semantic of belonging, and 510–540 other semantics

construct.[8]Semantic cohesion is identified with the capacity of each partner to use the characteristic meanings of the other in their own narrated story and to construct interactive polarities that belong to the same semantic in the here and now of the interaction. Our analysis makes it possible to assess both the overall semantic cohesion shown by the couple during the session, as well as the contribution of each partner to its construction. This cohesion is different and complementary to that put forward by the circumplex model of marital systems (Olson & Gorall, 2003; Olson, Sprenkle, & Russell, 1979).

These two variables make it possible to evaluate relational dynamics that go beyond the verbalized conflict, to identify the roots of the process of constructing meanings, and to explain the misunderstandings so often present between partners. They are explicative variables of the couple dynamics and conflicts rather than descriptive like most of the variables used in research on multiactor sessions. The proposed method also opens up the world of meanings for the research of multiactor sessions. Meaning, up to now analyzed by instruments aimed at grasping its individual processes,[9] can now be explored in multiactor settings catching the processes of its joint construction.

Identification of the couple's semantics and the assessment of the semantic cohesion also contribute to evaluate the quality of the dialogue (Seikkula, 2011) and provide a guide for the therapeutic process. It suggests specific strategies to the therapist in tune with the therapeutic relationship, which varies according to the partners' semantics (Ugazio, 2012, 2013).

Due to the lack of space, we could not examine how Victoria and Alfonso built up the relationship with their therapists. The study of the semantic exchange between the therapist and the couple is, however, a fundamental part of the method (Ugazio & Castelli, 2015).

The method is essentially qualitative with inferential aspects that allow us to overcome the main limitation of computer aided coding systems (CAQDAS), making it possible to identify meanings by inferring them from the context. Though qualitative, it allows for quantification useful for research purposes. However, it is a time-consuming method, a limitation that can be overcome by restricting the analysis to parts of the sessions, but which parts to submit to it becomes a crucial choice not without risks. The application of the method also requires a full in-depth grasping of the theoretical model underpinning it (as developed by Ugazio, 1998, 2012, 2013) and extensive training. In addition, the identification of the interactive polarities in multiactor therapeutic sessions necessitates clinical competence and experience in family therapy. These competences are not required for extracting the narrated polarities.

The construction of meaning is an essential component in the couple's relational dynamics, responsible for a significant part of their conflict and hardship, but other

---

[8] See section "Construction of Meaning Within a Couple Relationship".

[9] Such as the repertory grid (Kelly, 1955) and the semantic differential (Osgood, Suci, & Tannenbaum, 1957).

aspects of the couple's life are equally important. We are well aware of this. The proposed method is therefore complementary to other methods focused on different aspects of the couple dynamics.

References

Anderson, S. A., & Bagarozzi, D. A. (Eds.). (1988). *Family myths*. New York: Haworth.

Cronen, V. E., Johnson, K., & Lannamann, M. (1982). Paradoxes, double binds and reflexive loops: An alternative theoretical perspective. *Family Process, 20*, 91–112.

Ferreira, A. (1963). Family myths and homeostasis. *Archives of General Psychiatry, 9*, 457–463.

Guidano, V. F. (1987). *Complexity of the self*. New York: Guilford Press.

Guidano, V.F. (1991). *The Self in process: Toward a post-rationalist cognitive therapy*. New York: Guilford Press.

Guidano, V. F., & Liotti, G. (1983). *Cognitive processes and emotional disorders*. New York: Guilford Press.

Hermans, H. J. M., & Di Maggio, G. (Eds.). (2004). *The dialogical self in psychotherapy*. London: Brunner-Routledge.

Hinde, R. A., & Herrmann, J. (1977). Frequencies, durations, derived measures and their correlations in studying dyadic and triadic relationships. In H. R. Schaffer (Ed.), *Studies in mother-infant interaction* (pp. 19–46). New York: Academic Press.

Jung, C. G. (1921/1971). *Psychologische Typen*. Zurich: Rascher [Tr. Psychological types, collected works, Vol. 6. Princeton, NJ: Princeton University Press].

Kelly, G. A. (1955). *The psychology of personal constructs*. New York, NY: Norton.

Minuchin, S., Nichols, M. P., & Lee, W. Y. (2007). *Assessing families and couples: From symptom to system*. Boston: Pearson, Allyn and Bacon.

Minuchin, S., Rosman, B., & Baker, L. (1978). *Psychosomatic families: Anorexia nervosa in context*. Cambridge, MA: Harward University Press.

Neimeyer, R. A., & Raskin, J. D. (2000). *Constructions of disorder: Meaning-making frameworks for psychotherapy*. Washington, DC: American Psychological Association.

Ogden, C.K. (1932). *Opposition: A Linguistic and Psychological Analysis*. London: Kagan.

Olson, D. H., & Gorall, D. M. (2003). Circumplex model of marital and family systems. In F. Walsh (Ed.), *Normal family process* (3rd ed., pp. 514–547). New York: Guilford.

Olson, D. H., Sprenkle, D. H., & Russell, C. S. (1979). Circumplex model of marital and family systems: I. Cohesion and adaptability dimensions, family types and clinical application. *Family Process, 18*, 3–28.

Osgood, C. E., Suci, G., & Tannenbaum, P. (1957). *The measurement of meaning*. Urbana, IL: University of Illinois Press.

Procter, H. G. (1981). Family construct psychology: An approach to understanding and treating families. In S. Walrond-Skinner (Ed.), *Developments in family therapy* (pp. 350–366). London: Routledge & Kegan Paul.

Procter, H. G. (1996). Family construct psychology. In S. Walround-Skinner (Ed.), *Developments in family therapy* (pp. 350–356). London: Routledge.

Reiss, D. (1981). *The family's construction of reality*. Harvard, MA: Harvard University Press.

Rober, P. (2002). Constructive hypothesizing, dialogic understanding and the therapist's inner conversation: Some ideas about knowing and not knowing in the family therapy session. *Journal of Marital and Family Therapy, 28*, 467–478.

Seikkula, J. (2011). Becoming dialogical: Psychotherapy or a way of life? *The Australian and New Zealand Journal of Family Therapy, 32* (3), 179–193.

Sluzki, C. E. (1992). Transformations: A blueprint for narrative changes in therapy. *Family Process, 31*, 217–230.

Ugazio, V. (1998, 2012, new extended and revised edition). *Storie permesse, storie proibite.* Torino: Bollati Boringhieri.

Ugazio, V. (2013). *Semantic polarities and psychopathologies in the family: Permitted and forbidden stories.* New York: Routledge.

Ugazio, V., & Castelli, D. (2015). The Semantics Grid of the Dyadic Therapeutic Relationship (SG-DTR). *TPM—Testing, Psychometrics, Methodology in Applied Psychology, 22*(1), 135-155.

Ugazio V., & Guarnieri S., (2016). The Family Semantic Grid (FSG) III. Couples and Families interactive polarities. *TPM. Testing, Psychometrics and Methodology in Applied Psychology* (in press)

Ugazio, V., Negri, A., & Fellin, L. (2015). Freedom, Goodness, Power, and Belonging: The Semantics of Phobic, Obsessive-Compulsive, Eating, and Mood Disorders. *Journal of Constructivist Psychology, 28(4), 293-315.*

Ugazio, V., Negri, A., Fellin, L., & Di Pasquale, R. (2009). The Family Semantics Grid (FSG). The narrated polarities. *TPM—Testing, Psychometrics, Methodology in Applied Psychology, 16*(4), 165–192.

Villegas, M. (1995). Psicopatologías de la libertad (I). La agorafobia o la constricción del espacio. *Revista de Psicoterapia, 21*, 17–40.

Watzlawick, P., Beavin, J. H., & Jackson, D. D. (1967). *Pragmatics of human communication.* New York: Norton.

White, M., & Epston, D. (1990). *Narrative means to therapeutic ends.* New York: Norton.

Chapter 10
Constructing the Moral Order of a Relationship in Couples Therapy

Jarl Wahlström

Couples therapy is commonly seen as one modality of psychotherapy. Within that framework, the objects of treatment are understood to belong to the domain of psychology, however, it may be defined (Crowe, 1996). From a psychodynamic point of view, the objects of treatment involve the inner worlds of the partners and their mutual inter-dependencies. Cognitive–behavioral approaches seek to alleviate limitations in communication skills between spouses. Systemic therapies address dysfunctional patterns of interaction in the relationship. Through the perspective of attachment theory, emotionally focused marital therapy (Johnson, 2004) seeks to help clients explore and better manage their emotional experiences.

From the social constructionist (Burr, 1995; Gergen, 1994) and postpsychological (McLeod, 1997) points of view adopted in this study, merely psychological formulations of the goals and practices of couples therapy appear to be restricted in the sense that they do not take into account the institutionally framed constructive work of the spouses (Kurri & Wahlström, 2003). This is not to say that the psychological perspective would not be relevant. However, if the couples therapist's self-understanding of his or her professional activities is solely based on psychological theory, he or she will be naïve in respect to other salient aspects of the process. These have to do with how, in therapeutic conversations, the couple's relationship is presented and performed as a social institution with a particular social and moral order.

In this case study, it will be asked how the discursive practices of the participants in a couples therapy process establish the sessions as an arena for constructing the moral order of the relationship of the partners. The aim is to show, through a detailed discursive analysis of four illustrative episodes in the process, how positionings and meaning-constructions relevant to forming a moral order are performed and how the essential moral dilemmas of this particulate case can be formulated.

J. Wahlström (✉)
Department of Psychology, University of Jyväskylä, Jyväskylä, Finland
e-mail: jarl.wahlstrom@jyu.fi

© Springer International Publishing Switzerland 2016
M. Borcsa, P. Rober (eds.), *Research Perspectives in Couple Therapy*,
DOI 10.1007/978-3-319-23306-2_10

The study is based on four sessions of therapy with two relatively young adults, Victoria and Alfonso, from two different cultural environments (northern and southern Europe) and in an early phase of their relationship.

Theoretical Introduction

As Humberto Maturana (1988) has put it, people enter couple relationships driven by "a passion for living together in close proximity" (Efran, Lukens, & Lukens, 1990, 158). This view of humans as biologically tuned towards relationships has been endorsed by contemporary affective neuroscience (Panksepp, 2009). The word "love" is commonly used to denote this fitting together, remaining together, and continuous remaking of interactional patterns, in which people as biological (and psychological) systems become involved with one another. Love takes place without prior justification and there is no "reason" for it. In this sense, love as a passion is blind. It does not inform people about how to do "the living together." Still, as Kenny (1985) remarks, love as a primary constitutive condition is also fundamental for social phenomena. This brings us to the question of the social and moral order of the couple's relationship and the intricate relationship between emotions on one hand and actions based on moral judgments on the other—a territory often left unexplored by theorists and practitioners of couples therapy.

A Joint Form of Life and the Moral Order

After entering a relationship, couples have to construct a joint form of life (Kurri & Wahlström, 2003), which means establishing their relationship as one particular instance of a social institution. This process involves applying, within mundane activities, such social practices as mutual positioning (Davies & Harré, 1990) of self and other, and negotiating criteria for diverse category memberships (Widdicombe, 1998), particularly those of being a "wife," a "husband," a "spouse," or a "partner." In (post)modern society, few opportunities remain for one to rely on traditional practices and rituals when doing such constructive work. Consequently, "negotiations" play an increasingly central part in this process, which includes as an essential element creating the social and moral order (Harré, 1983) of the relationship. The moral order of a relationship includes more or less articulated and shared understandings of what is valued and what is not; the loyalties, duties, and responsibilities expected from the partners and the grounds for evaluating actions. It also includes expectations of how value, concern, and respect are communicated.

Couples therapy can be seen as a special, and in some sense privileged, arena for such kinds of "negotiations," a perspective mostly overlooked in theory and research. So-called "negotiating" is done indirectly and it involves discursive practices that fulfil manifold communicative tasks, such as problem formulations, clienthood construction, blame and counter-blame, complimenting and giving credit, arguing and reconciling, and others. All these discursive practices unfold in therapeutic conversations without the participants' conscious intention to formulate or negotiate any moral or ethical principles. Therefore, the constructive work in this realm is implicit and usually remains "invisible" (Kurri, 2005) for the participants themselves. Accordingly, it has to be recognized within the flow of conversation through interpretational efforts.

There are two good reasons for researching how a moral order is constructed in couples therapy talk. First, even though the conversational acts that establish a moral order can be observed in the interaction, explicating their function as constituents of that order calls for a theoretically grounded analytic reading. Secondly, the concept of a moral order is closely connected to two other important theoretical concepts, namely agency and positioning. These concepts are also of utmost practical importance for any therapeutic enterprise. Establishing new and more agentic positionings for clients in respect to self, others and problems can be seen as the core process in psychotherapy (Avdi, 2012; Leiman, 2012; Wahlström, 1990, 2006a, 2006b).

Positioning

Social encounters are not just meetings between individuals, but between people undertaking particular social commitments. In the social sciences, this has been framed by the concept of role (Suoninen & Wahlström, 2009). When performing a role, an individual conforms to others' expectations regarding his or her behavior in a certain situation. For instance, compared to a client's role, a therapist's role involves different expectations of what the situation requires from the individual. However, actual interaction is hardly ever merely a ritual of performing role expectations and role-based descriptions of institutional interaction consequently miss much of the richness of situational performance. Within the same basic role staging, participants create a variety of interactional settings.

The concept of position (Avdi, 2012; Davies & Harré, 1990), which suggests a more flexible notion of social staging than the concept of role, seeks to account for this situational variability. A position is always interactional, taken in respect to something or somebody else and thus suggests positions for others, hence the action-term 'positioning'. When different positionings emerge in an institutional meeting, the evolving conversational setting affects what the situation can afford (Suoninen & Wahlström, 2009). With changing positions, speakers will vary the accounts they give of events and the descriptions they share of the characteristics, rights, and duties attributable to those involved. Positions appear in different combinations (i.e., including positioning of both self and others) and they are essential constitutive elements of the emerging social and moral order.

Problem Formulations and Clienthood Negotiations

Problem formulations in therapy are not neutral (Buttny, 1996; Buttny & Jensen, 1995; Kurri & Wahlström, 2005). Usually a problem formulation takes the form of a description of an undesirable state of affairs and thus invites a process of change. But there is more to it than just that. Deliberations on the question "what is the problem?" will eventually turn towards the questions "who is responsible for solving it?" and "who is the one who should change?," thereby necessarily committing the speakers to one or another moral understanding. Giving an account of problems includes placing or taking responsibility and consequently it constitutes an act of executing moral judgement.

This is closely connected to the question "who is the client?." It is not uncommon in couples therapy for the position of client to be assigned by spouses to each other and the therapist is called upon to work on the problems of one on behalf of the other (Kurri & Wahlström, 2003). Because of this, "negotiations" concerning clienthood are at the core of the therapeutic process. To whom and how the position of "client" is assigned fundamentally influences how the manifold issues in the couple's life will be dealt with in the sessions. Problem formulations and negotiations of clienthood not only take place at the beginning of the therapy, but are also, albeit often only implicitly, present throughout the entire process.

Blame and Accountability

Problem formulations in couples therapy are frequently expressed in the form of blame (Buttny, 1990, 2004; Kurri & Wahlström, 2003, 2005). Blaming is a formulation of an unwanted state of affairs in which the responsibility of causing, and potentially also repairing, the situation is usually put on somebody else than the speaker. In instances of self-blame, the responsibility is given to the speaker him or herself. While blame constructions in therapy discussion can be explicit and direct, they are often complicated and implicit, including indirect linguistic formulations and nonverbal and paralinguistic clues. The blame can take the form of a seemingly neutral description of a situation or a person, and its discursive status as a blame is seldom unequivocal.

Being a target of blame puts the blamed person in a morally vulnerable position that threatens his or her social "face" (Goffman, 1971). This calls for remedial work on his or her part, by means of which the meaning of a presumably unacceptable or offensive act is improved to the point that it is acceptable. To defend his or her moral status in the conversation, the targeted person may respond by counter-blame. More often, however, the person being blamed finds him or herself compelled to give an account of his or her behavior. Scott and Lyman (1968) define an account as a linguistic device that is employed whenever someone's action is subjected to evaluative inquiry and they specify two different types of accounts: excuses and justifications.

In justifications, the speaker accepts responsibility for an act, but denies its negative quality. Excuses are accounts in which the speaker admits the reprehensible character of an act, but denies full responsibility for it.

Blame constructions in couples therapy are sensitive and delicate discursive moves (Kurri & Wahlström, 2003, 2005). Because the person who blames is also in a position needing accountability, he or she has to create and justify his or her moral status as a person who is entitled to assign blame. This is often seen as softened expressions and circumstantial formulations in blaming utterances. Taking a blaming position in a couple relationship potentially makes the speaker morally vulnerable, since blame can easily reflect back on him or her: why did the speaker choose to form a relationship with such a reprehensible person? Blame and counter-blame sequences are informative from a therapeutic point of view, as they concern the relational patterns of the couple.

Agency

Positions within social interaction—taken or given—can be more or less agentic. Core features of human agency include intentional causality and conscious understanding thereof, as well as the capacity to distinguish between self-caused and externally caused phenomena (Kögler, 2012). Agency requires a reflexive relation towards oneself enabled by an external viewpoint (i.e., a position in which a person takes the perspective of others to reflect on him or herself) (Gillespie, 2012). From a social constructionist perspective, taking an agentic position corresponds to participating in conversations that produce meaning for the person's life (Drewery, 2005). Depending on the conversation, some positions are more agentic (i.e., constructing for the person an active and responsible stance), while others are more nonagentic (i.e., reducing the person's possibilities to influence his or her situations and actions).

Earlier research on psychotherapy talk has shown that disclaiming one's own agency is, in fact, a relatively common discursive practice adopted by clients (Kurri & Wahlström, 2001, 2005, 2007; Partanen & Wahlström, 2003; Partanen, Wahlström, & Holma, 2006; Seilonen, Wahlström, & Aaltonen, 2012). Detailed case studies of therapy discussions have revealed a variety of discursive means used by clients to achieve the conversational goal of actively presenting oneself as nonagentic in relation to the events in one's life. This is in itself an agentic act, however, which results in the simultaneous use of different displays of agency or "split agency," in which the self can be presented as an active and responsible participant in the actual therapeutic situation and, at the same time, as a weak or "acratic" agent in life events. Such multiple presentations of self as agent serve different functions connected to establishing and sustaining a viable moral order within the session (i.e., managing the distribution of rights and duties, accountability and responsibilities, as well as preserving the moral face of participants).

Data and Analysis

The data of this study is comprised of talk produced by two clients and two therapists in their conversations during four couples therapy sessions. This data was made available for the researcher in the form of video recordings and transcriptions of these recordings. The transcripts are verbatim with some special notations indicating prosodic and interactional features of the talk.

Methodological Approach of Analysis

This research on the conversational construction of a moral order includes two methodological points of view. The first one is the construction of positions and participation frameworks within relevant speech actions (Goffman, 1981) (i.e., the formal side of the interaction), and it is guided by ideas generated from conversation analysis (CA) (Peräkylä, Antaki, Vehviläinen, & Leudar, 2008). The second one is the construction of meaning within the relevant passages. This means looking at the content of language use and the meaning-worlds that are constructed in the conversation. These are questions typically addressed by different approaches within discourse analysis (DA) (Potter, 2004; Wetherell, 2001), discursive psychology (DP) (Edwards & Potter, 1992; Harré & Stearns, 1995; Potter, 2003), and social constructionism (Burr, 1995; Gergen, 1994).

The key concept of analysis is that of the moral order, which is seen as being constantly constructed in conversational interaction. The moral order as such cannot be observed. However, the construction of the moral order can be observed as conversational acts performed by the participants. As mentioned in the introduction, such acts include blaming, complimenting, judging, prescribing actions, defining rights, duties and loyalties, and so on. In actuality, it is any act which contributes to how value is defined and distributed in the conversation. Hence, these kinds of speech acts were the primary units of analysis.

Analytic Procedure

The aim of the analysis was to reconstruct how in their constructive work the participants offered suggestions for the moral order of the clients' relationship. This meant trying to find answers to three questions: (1) *What* contents relevant to the construction of the moral order of the relationship were present in the data? (2) *How* did the participants perform the construction of the moral order of the relationship in their utterances and speech acts? (3) *Why* was the moral order of the relationship constructed as it was?

To answer the first question ("*What* contents relevant to the construction of the moral order of the relationship were present in the data?"), a thematic analysis of the data was done. First, the sessions were partitioned into segments on the basis of the main conversational agendas pursued by the participants. The segments were given headings reflecting the researcher's understanding of the agenda at hand. The identification of segments and the assignment of headings were the only interpretational aspects of the thematic analysis. The contents listed under the headings were merely condensed presentations of the primary data and they helped to navigate within the data corpus.

Then, secondly, based on several readings of the transcribed text and repeated watching of the video recordings, the key meaning-constructions and positionings performed by the speakers were indicated and listed for each topical segment. Notes were made as to the agentic status of the positions taken by the speakers. Analytic ideas derived from positioning theory (Harré & Van Lagenhove, 1999) and from participation theory (Goodwin, 2007), as well as from earlier research, were used here.

To answer the second question ("*How* did the participants perform the construction of the moral order of the relationship in their utterances and speech acts?"), four episodes from different phases of the therapeutic process were chosen for a detailed, turn-by-turn discursive analysis. The selection of these episodes was based on the global thematic reading of the data, and they were judged to be representative of the discursive practices in use and the development of positionings within the emerging moral order, as it was observed throughout the therapy. Analytic principles and conceptual tools from CA, DA, DP, and social constructionism (see above) were used in this reading with the aim of giving a detailed description of how positionings and meaning-constructions were performed. The findings of this analysis constitute the core of the Results part of this study. Extracts from the primary data are also shown, giving the reader the possibility to evaluate the credibility of the analysis.

The answer to the third question ("*Why* was the moral order of the relationship constructed as it was?") can be found in the Discussion part of this study. It is presented as a summary and more general conclusion of the findings, and as such it constitutes the researcher's statement, open to further debate by readers.

Results

The results of the study will be presented as detailed analyses of four exchanges of conversational turns, each of them from different sessions. The first exchange from the first session shows some aspects of the initial problem formulation. The extract from the second session shows one of many episodes in which Alfonso's relationship to his family of origin and the misgivings that Victoria had in respect to this were discussed. The third extract from the third session is part of a longer segment in which the participants returned to the initial problem formulation by exploring in detail a recurrent problematic pattern in the couple's relationship. The extract from the final session concerns a conversation on an issue in connection to their present life situation (i.e., moving to a new apartment). Each extract is analyzed keeping in mind the research question on how the moral order of the relationship and the agencies of the spouses are constructed in the conversation.

Formulating the Problem

At the very beginning of the first session, in response to the therapist's question "So where would you start?," Victoria referred to her depression: "The reason why we are here is that this, my thing, left scars in our relationship." The first extract shows how the situation was discussed a few minutes later.

Extract 1

Session 1, turns 24–29

24 V   yes, I needed only some tiny reason and I made it grow and grow (.) and I didn't trust him in anything, I didn't trust myself and I didn't trust him, it was like I made everything grow in such a huge problem in my head (.) now I don't do it any more (.) but I really, like, I still need to talk a lot about everything, like if there is anything, I just need to solve it right there but I feel like now Alfonso is not able anymore because he's afraid

25 A   yeah, like kind of, that I just can't

26 T   kind of

27 A   I kind of feel like I can't deal, like I, I before, I felt like I had all this, somehow, patience to listen and, even if it was like for a long, for a long time, this kind of situation now I kind of feel that it's, for whatever small thing that I feel that I get like

28 V   you get in panic, somehow, very anxious like somehow

29 A   yes, it's like, yes (.) I think it's kind of I get afraid that it could be again some similar situation

In the first part of the 24th turn of the conversation (the first turn of this extract), Victoria describes a situation in which she was depressed. Here she exhibits a typical instance of "split" agency. On one hand, she takes responsibility for her action ("yes, I needed only some tiny reason," "I didn't trust"). On the other, she describes herself as nonagentic in the situation. Her past behavior becomes justified as a manifestation of her psychological condition at that time. When she now—in the present conversational context—exhibits a reflexive stance in respect to her previous behavior, she establishes for herself a position as a trustworthy conversationalist.

Secondly, Victoria makes a distinction between her past and present ways of acting ("now I don't do it any more (.) but I really ... still need to talk a lot about everything"). These formulations create a position from which she can defend her moral status in the present conversation. The change in self-categorization from a depressed person to a person who is past depression gives different grounds to her claims. The justification for her plea is still psychological, though. It is her "need" to talk, not her "wish" or her "demand." For the partner, to refuse a need is a morally questionable act, which is quite different from refusing a wish or even more so a demand. A "need" implies some restriction of agency on the part of the speaker, and hence the potential responsibility of a partner to accommodate that "need."

From this position, Victoria assigns blame that precisely addresses a failure to accommodate her need ("but I feel like now Alfonso is not able anymore"). She softens the blame by providing an excuse for Alfonso's undesirable behavior ("because he's afraid"). By doing this, she creates for herself a still stronger agentic position in the conversation as the person who gives meaning, not only to her own but also her partner's behavior. At the same time, the nonagentic position in respect to the problem becomes shared. Both partners are put in a position of not being able to act as they presumably would want to: Victoria because of her "need," Alfonso because of being "afraid." In this way, an instance of "trouble-talk" is performed, which creates an appropriate starting point for a therapeutic conversation, as well as a suggestion for how it should be focused.

There is immediate uptake on Alfonso's part. He partially accepts the blame ("yeah, like kind of, that I just can't"), but proceeds to qualify this acceptance. By stating "I kind of feel like I can't deal, like I (did) before," he justifies his position. Earlier he had been acting as Victoria wished ("I had all this … patience to … listen"), but now he is not capable of doing that anymore. Victoria and Alfonso together construct a justification for this ("you get in panic" and "yes … I get afraid"), and thus this initial problem formulation creates a conversational situation where both partners are positioned as powerless victims of psychological forces (her "need," his "panic"). But at the same time a potential moral dilemma is presented: is Alfonso obligated on the basis of some moral grounds to overcome his "fear" and respond to Victoria's "need"? Is Victoria likewise obligated to take into account Alfonso's "fear" and disallow her "need"?

Weighing Loyalties

In the second session, after having explored the couple's present situation and the consequences of a task that was given to the clients in the first session, the therapist asked "What would you like to do this time here? How would you like to use this time?" Alfonso responded by saying that he is going to visit his home country and his family, and that "it will be like, good to see how that turns out." It has been discussed how Victoria feels that Alfonso forgets her when he is visiting his home country. Victoria has said about Alfonso "that he doesn't think about me, that he kind of likes to forget about me, when he is there." She has wished that he would send her SMS messages during his stays abroad, but has also stated that "it should come a bit naturally." The second extract shows a piece of the discussion on this theme.

Extract 2

Session 2, turns 241–248

241 A   yes, so I think it's not the same situation, we are in two different places, we are apart, she's thinking about, still I think I'm in a different context

242 T   it seems a bit, it seems to have, it seems to have become a big issue this and a very concrete detail in your relationship

243 V   yeah (.) and, yeah it is a big thing because, I know that family is important, and I also (.) I have tried also to I don't want Alfonso to be between two families

244 T   mm

245 V   I think I have tried, but then I feel do they want to keep you so busy so that (.) you don't have time for me, because it's clear that they don't like me

245 A   I think it's

246 V   it's a difficult situation, and I don't know if, Alfonso says that I am important and I really don't want to put him in this a situation, but I am not sure that his family doesn't want to put him in that situation

247 T   which kind of situation?

248 V   that he has to be between two fires

In turn 241, Alfonso defends his position by accentuating the difference between his and Victoria's situations ("I think it's not the same situation" and "I'm in a different context"). Here he is indirectly pleading for his right to be considered in light of his circumstances (i.e., for Victoria to back off from her request, taking his situation into account). This plea is accentuated in his rhetoric when he acknowledges Victoria's stance ("we are apart, she's thinking").

The therapist designates the topic as important ("it seems to have become a big issue and very concrete") and thus worthy of being dealt with in the conversation and then makes an important categorization by defining Victoria's and Alfonso's situation as a "relationship." In her response to this, Victoria makes significant constructive work. She voices a general principle ("I know that family is important"). In doing so, she shows herself as being capable of taking other points of view, including that of Alfonso, into consideration. Therefore, her claim cannot be dismissed on grounds of a lack of concern for Alfonso. Secondly, she redefines the "relationship," giving it a higher institutional status ("I don't want Alfonso to be between two families"). By using the word "family" to define her and Alfonso's relationship, Victoria justifies her institutional rights as equal to those of his family members.

This also works as a ground for putting the blame for creating the conflict between the two "families" on Alfonso's family ("I think I have tried, but then I feel do they want to keep you so busy so that you don't have time for me"). The position of Alfonso's family is constructed as unambiguous ("because it's clear that they don't like me"), which serves to reduce either her or Alfonso's responsibility for the conflict. In Victoria's formulation, however, some uncertainty still remains regarding Alfonso's stance ("I don't know if Alfonso says that I am important"). This expression appears to call for some accountability on his part. In Victoria's version, Alfonso "has to be between two fires," a vivid metaphor for a loyalty conflict, but she evades her responsibility of being one of the actors who has put him there.

The dilemma of the moral order of the relationship created here can be formulated as follows: Is Victoria justified in her demands that Alfonso demonstrate her
presence in his mind when visiting his family of origin? Is Alfonso justified in his
demands that Victoria consider his sensitive situation when visiting his family and
give up some of her requests?

Exploring a Core Conflict Pattern

In the third session, when the therapist asked "what would you like to do today, to
use this time?," Victoria responded by saying: "I think that, that things are fine now
only the same thing that, why I contacted you in the first place is that he gets, so
scared...." She then related one incident when she had asked if she could come and
listen to a musical event where Alfonso was playing. He said yes, but a few days
later Victoria had asked "do you even want me to come?". She justified this question
in the session by saying "because I was the one who invited myself, really not with
any deeper meaning, just a simple question, I didn't mean anything with it, I just
asked if he wants me to come." This led to a fight between the couple. This incident,
and especially Alfonso's reaction to Victoria's question, was discussed at length in
the session. Extract 3 is an exchange of turns in that discussion.

Extract 3

Session 3, turns 141–149

141 A	Maybe after that we, after these things turned like this, after that I started having this kind of reaction I think, I think maybe not, maybe at some level, maybe sometimes, sometimes a bit more, sometimes a bit higher, sometimes lower level, but I always have this
142 T	Mmm(.) And how would you define the questions that make this happen? I suppose that there are many questions to each other during the day that
143 A	I think it's some questions about, they are this kind of questions like (.) kind of (.) how to say? (.) maybe when I have to (.) explain something like (.) or prove (.) prove (.) something to her like that
144 V	like usually it's really some simple question that I would need like one word for an answer, but then I don't get it, I get only this awful, like (.) this very bad reaction
145 T2	What kind of a reaction those are? What do you mean by that?
147 V	Alfonso's reaction is like, his face gets like this and like, I don't know, I think I have explained it but I don't know if you were here (.) but he gets like really suffering (..)
148 A	yeah, it's a bit like, when you are kind of disappointed, you are a bit down, a bit
149 V	and then for very small reasons I think this happen like, like I think that in every relationship there is times that you, you want to talk about your relationship, it doesn't work like if you never talk about it, and even if I try to talk about positive things (.) for example once I remember I asked you something that was, I meant it to be a positive thing, but you immediately thought that I had some intentions so like that I have a deeper hidden meaning that I want to get you in a trap, and then it happened again, like he doesn't trust me, he thinks that I always just try to (.) I don't know, I don't know how to explain it but I just feel like I can't talk about things and I don't have the right to feel sad any more or disappointed or anything, that if I need to talk about something like commonly, normally, positive or negative things I feel like we are not able, anymore, and it's very frustrating and we really need to get past this

In turn 141, Alfonso seeks to find a way of expressing his difficulties of responding to Victoria's questions. The use of mitigating formulations ("sometimes maybe," "this kind of," "a bit more," "a bit higher") and excessive repetition indicates that he is approaching the topic as a delicate and sensitive one. He refers rather vaguely to "this kind of reaction" "I started having,", offering his behavior as the topic that should be focused on. In his turn, the therapist prompts Alfonso to define "the question" more precisely, which directs the focus towards the interactional pattern as a whole, as well as Victoria's part in it. Still in a very sensitive mode, Alfonso points to the difficulties he experiences when confronted with her questions. He does not explicate what it is in Victoria's behavior that he finds difficult to cope with, but it appears that this might be his sense of her mistrust.

Although Alfonso does not explicitly blame Victoria, her response takes the form of counter-blame. As is typical of a blame construction, she refers to the undesirable behavior as being recurrent ("usually") and undue as a reaction to her request, as well as to what it would entail ("it's really some simple question"). Again she justifies her request as stemming from her "need", which serves to make it difficult to refute without compromising the speaker's own moral status. Finally, the blame construction is given force by the use of a strong formulation ("this awful ... this very bad reaction"). Victoria's turn serves to direct the topic of the conversation to Alfonso's behavior (T2: "What kind of a reaction those are? What do you mean by that?").

In turn 149, Victoria gives an account of the couple's situation that is cast in the form of a complaint or blame. First, she marks what she is going to say as important by giving it a generic reference ("in every relationship" and "it doesn't work ... if you never talk about your relationship"). Then the rhetoric of the turn is strengthened by giving an example ("for example, once I remember"). This example works as an exception that reinforces the rule ("I meant it to be a positive thing"), and it presents the rule that patterns the relationship as undeniable. In Victoria's account, the pattern is that because of Alfonso's misreading of her intentions ("that I want to get you in a trap"), there is no longer any room to talk about things (i.e., negotiate the relationship).

From the point of view of the moral ordering of the relationship, it is significant that in her complaint Victoria refers to the couple's predicament as a loss of her rights ("I don't have the right to feel sad anymore or disappointed or anything"). This can be seen as a consequence of the earlier expressed rule "it doesn't work if you never talk about your relationship," which now can be read to be meant as not only descriptive but also prescriptive (i.e., having a moral bearing). In this specific context, the complaint expressed by Victoria also works as a bid for a goal and an agenda for the therapeutic work (i.e., this is something "we really need to get past"). Here the responsibility for change is given to both partners and the sense of a lost agency is presented as shared. Working on the relationship is presented as a moral obligation, a value statement that is difficult to refute within the couples therapy context.

Differing Commitments

In the fourth and last session, there is a discussion about the couple's plans for the near future. It turns out that they have moved into a new apartment, which means that they have had to—and still have to—make decisions about furniture, decorating, carpets, etc. The fourth extract shows a brief sequence of turns from this discussion.

Extract 4

Session 4, turns 180–189

180 V	no but that's just because Alfonso doesn't like it, like right now we don't have any carpets because in [Alfonso's home country] they don't have carpets he doesn't like them, but to me it doesn't feel like home when there's no carpets
181 T	so it actually is a big issue, yeah (.)
182 A	yeah in the end, that's also, it's not so (.) it's not so challenging (.) we'll find some way
183 T	why, why isn't it so challenging, what do you think?
184 A	because maybe sometimes (.) we were having a move, when we moved to this last place, that I was just (.) for example to me (.) a house it's OK, like it's not the most important (.) like for her it's really important
185 V	what?
186 A	the house, like this feels home and this kind of thing
187 V	mm
188 A	but for me not so much or may be to me maybe, some example what could be, if we have to, for example we move to this new place and maybe I don't think that we should buy some new stuff, to me, it's, it's, I think maybe it's just not so important to me
189 V	but for me home is like the most important thing (.)

In turn 180, the issue about having carpets or not is constructed by Victoria as cultural ("right now we don't have any carpets because in Alfonso's home country they don't have carpets"), but also as very personal ("he doesn't like them, but to me it doesn't feel like home when there's no carpets"). The therapist follows by marking the topic as important ("so it actually is a big issue"). Alfonso mitigates the importance of the issue ("it's not so challenging … we'll find some way"), but when prompted by the therapist ("why isn't it so challenging"), he acknowledges a crucial difference in the partners' attitudes ("to me a house … it's not the most important … like for her it's really important … like this feels home and this kind of thing … maybe it's just not so important to me"). Victoria responds to this by making a strong statement, "but for me home is like the most important thing."

In this exchange, the seemingly mundane issue of having carpets or not acquires important metaphorical meaning. For Victoria, having carpets means furnishing a home—"the most important thing"—while Alfonso, even while acknowledging Victoria's stance on the issue, makes it clear that living in a place that "feels like home" is not a high priority for him. There is a clear indication of differences of commitment to the relationship between the partners in these formulations. It is notable that in spite of this, the topic is not expanded on in the session, which ends soon after this exchange, even ahead of schedule. Thus, the difference in commit-ment is not brought up as an issue in the therapeutic agenda.

Discussion and Conclusions

In this case study, a detailed turn-to-turn discursive analysis of four conversational episodes from four sessions of couples therapy was performed. The episodes, one from each session, were chosen on the basis of how well they represented and illus-trated salient features of how moral dilemmas were presented and dealt with in the data as a whole. The aim of the analysis was to achieve some understanding of how the participants in the therapeutic conversations constructed the moral order of the couple's relationship. The selected extracts from the data were judged to be repre-sentative of the total text, both on thematic (what?) and procedural (how?) levels.

The continuous and open-ended process of constructing a moral order was con-ceived of as the production of utterances, where meanings were given to what was valued in the relationship and what was not, in addition to the loyalties, duties, and responsibilities expected of the partners and the grounds for evaluating actions. In the first episode, a moral dilemma was established concerning Alfonso's eventual responsibility to accommodate his spontaneous emotional reaction (phrased as "fear") and response to Victoria's wish (phrased as "need") to deliberate on her concerns over the relationship. The dilemma created in the second episode could be formulated as the question of whether or not Victoria was justified to claim a posi-tion in Alfonso's life that was equivalent (or even more) to that of his family of origin. In the third episode, Victoria sought to prescribe a generic rule of interrelat-edness ("it doesn't work if you never talk about your relationship") for the present relationship, again justifying this on grounds of her emotional needs. In the fourth episode, a mundane question concerning home furnishing (having carpets or not) was constructed as having both cultural and personal significance and, accordingly, being indicative of the partners' commitment to the relationship.

Arriving at an answer for the third question of this study ("*Why* was the moral order of the relationship constructed as it was?") can be attempted on two levels. The first has to do with the relationship of the clients. A common component of the moral dilemmas exhibited was how *relationality* on one hand and *autonomy* on the other should be valued. Relationality as a moral value was mostly pre-sented by Victoria, while autonomy was presented by Alfonso. Initially Victoria justified her position by presenting herself as depressed (i.e., weak and needy). This limited Alfonso's possibility of defending autonomy as a moral value. When the focus of the conversation moved from the dyadic relationship to the relation-ship with families of origin, including Alfonso's conflicting loyalties, the relationality vs. autonomy dilemma was reframed. He found a new position from which he could defend his autonomy discourse, and Victoria was able to show some understanding of it. She could articulate her hopes and wishes for the relationship from a more agentic position than that of a depressed person. And Alfonso could express his wish for independence more freely without being put in the position of an unconcerned person.

All of this opened a potential space for discussing the commitment of the partners. When watching the videos and reading the transcripts of the conversations in hindsight, as one is privileged to do in a study like this, it appears evident that Victoria and Alfonso had quite different commitments to their relationship. She was very much committed to building a family and a home, while he appeared to still be in a move from his family of origin and in a phase of transition towards independency in his life. The dilemmas and differences in negotiating the moral order of the relationship revealed by means of detailed turn-to-turn analysis make sense as manifestations of these essentially different life situations of the two young adults.

This leads to the second level of trying to answer the *Why?* question. Why was the issue of commitment not brought to the forefront of the therapeutic agenda? This clearly has to do with the initial problem formulation. The issue brought to the attention of both therapists and clients alike was a very psychological one: Victoria's depression and Alfonso's emotional difficulties in dealing with it. From a clinical point of view, it appears that much progress was reached with this issue, and the decision to end therapy was a mutual one. This was what was "visible" for the participants. However, the concurrently ongoing "negotiation" of the moral order of the relationship, as it could be made explicit in the present analysis, appears to have remained largely "invisible" for them, therapists, and clients alike.

What conclusions for the practice of couples therapy can be drawn from an analysis like this? It seems apt to formulate the answer to that question in the form of a moral dilemma: should couples therapists respect the problem formulations given by clients and the space of solutions that they imply, or are couples therapists, on moral grounds, obliged to use some generic understanding of relational problems to bring to the surface issues not defined by clients as problems to be worked on? This question has bearings on the agency of both therapists and clients, and it deserves due attention in debates on the self-understanding of couples therapy.

References

Avdi, E. (2012). Exploring the contribution of subject positioning to studying therapy as a dialogical enterprise. *International Journal for Dialogical Science, 6*, 61–79.

Burr, V. (1995). *An introduction to social constructionism*. London: Routledge.

Buttny, R. (1990). Blame-accounts sequences in therapy: The negotiation of relational meanings. *Semiotica, 78*, 219–247.

Buttny, R. (1996). Clients' and therapist's joint construction of the clients' problems. *Research on Language and Social Interaction, 29*, 125–153.

Buttny, R. (2004). *Talking problems: Studies on discursive construction*. Albany, NY: State University of New York Press.

Buttny, R., & Jensen, A. D. (1995). Telling problems in an initial family therapy session: The hierarchical organization of problem-talk. In G. H. Morris & R. J. Chenial (Eds.), *The talk of the clinic: Explorations in the analysis of medical and therapeutic discourse* (pp. 19–47). Hillsdale, NJ: Erlbaum.

Crowe, M. (1996). Couple therapy. In S. Bloch (Ed.), *An introduction to the psychotherapies*. Oxford: Oxford University Press.

Davies, B., & Harré, R. (1990). Positioning: The discursive production of selves. *Journal for the Theory of Social Behavior, 20*, 43–63.

Drewery, W. (2005). Why we should watch what we say. Position calls, everyday speech and the production of relational subjectivity. *Theory & Psychology, 15*, 305–324.

Edwards, D., & Potter, J. (1992). *Discursive psychology*. London: Sage.

Efran, N. S., Lukens, M. D., & Lukens, R. J. (1990). *Language, structure and change*. New York: Norton.

Gergen, K. J. (1994). *Realities and relationship: Soundings in social construction*. Cambridge: Harvard University Press.

Gillespie, A. (2012). Position exchange: The social development of agency. *New Ideas in Psychology, 30*, 32–46.

Goffman, E. (1971). *Relations in public: Microstudies of the public order*. New York: Basic Books.

Goffman, E. (1981). *Forms of talk*. Oxford: Blackwell.

Goodwin, C. (2007). Participation, stance and affect in the organization of activities. *Discourse and Society, 18*, 53–73.

Harré, R. (1983). *Personal being*. Cambridge, MA: Harvard University Press.

Harré, R., & Stearns, P. (Eds.). (1995). *Discursive psychology in practice*. London: Sage.

Harré, R., & Van Lagenhove, L. (Eds.). (1999). *Positioning theory*. Oxford: Blackwell.

Johnson, S. M. (2004). *The practice of emotionally focused marital therapy: Creating connection*. New York: Brunner-Routledge.

Kenny, V. (1985, October 2). *Life, the multiverse and everything: An introduction to the ideas of Humberto Maturana*. Invited paper presented at the Istituto di Psicologia, Universita Cattolica del Sacro Cuore. Retrieved from http://www.oikos.org/vinclife.ht

Kögler, H.-H. (2012). Agency and the other: On the intersubjective roots of self-identity. *New Ideas in Psychology, 30*, 47–64.

Kurri, K. (2005). *The invisible moral order: Agency, accountability and responsibility in therapy talk* (Jyväskylä studies in education, psychology and social research, 260). Jyväskylä: Jyväskylä University.

Kurri, K., & Wahlström, J. (2001). Dialogical management of morality in domestic violence counselling. *Feminism & Psychology, 11*, 187–208.

Kurri, K., & Wahlström, J. (2003). Negotiating clienthood and the moral order of a relationship in couple therapy. In C. Hall, K. Juhila, N. Parton, & T. Pösö (Eds.), *Constructing clienthood in social work and human services: Interactions, identities and practices* (pp. 62–79). London: Jessica Kingsley.

Kurri, K., & Wahlström, J. (2005). Placement of responsibility and moral reasoning in couple therapy. *Journal of Family Therapy, 27*, 352–369.

Kurri, K., & Wahlström, J. (2007). Reformulations of agentless talk in psychotherapy. *Text & Talk, 27*, 315–338.

Leiman, M. (2012). Dialogical sequence analysis in studying psychotherapeutic discourse. *International Journal for Dialogical Science, 6*, 123–147.

Maturana, H. (1988). Reality: The search for objectivity or the quest for a compelling argument. *Irish Journal of Psychology, 9*, 25–82.

McLeod, J. (1997). *Narrative and psychotherapy*. London: Sage.

Panksepp, J. (2009). Brain emotional systems and qualities of mental life. From animal models of affect to implications for psychotherapeutics. In D. Fosha, D. J. Siegel, & M. F. Solomon (Eds.), *The healing power of emotion* (pp. 1–26). New York: Norton.

Partanen, T., & Wahlström, J. (2003). The dilemma of victim positioning in group therapy for abusive men. In C. Hall, K. Juhila, N. Parton, & T. Pösö (Eds.), *Constructing clienthood in social work and human services: Interactions, identities and practices* (pp. 129–144). London: Jessica Kingsley.

Partanen, T., Wahlström, J., & Holma, J. (2006). Loss of self-control as excuse in group-therapy conversations for intimately violent men. *Communication & Medicine, 3*, 171–183.

Peräkylä, A., Antaki, C., Vehviläinen, S., & Leudar, I. (Eds.). (2008). *Conversation analysis and psychotherapy*. Cambridge: Cambridge University Press.

Potter, J. (2003). Discursive psychology: Between method and paradigm. *Discourse & Society, 14*, 783–794.

Potter, J. (2004). Discourse analysis as a way of analyzing naturally occurring talk. In D. Silverman (Ed.), *Qualitative research: Theory, method and practice* (2nd ed., pp. 200–221). London: Sage.

Scott, M., & Lyman, S. (1968). Accounts. *American Sociological Review, 33*, 42–62.

Seilonen, M.-L., Wahlström, J., & Aaltonen, J. (2012). Agency displays in stories of drunk driving: Subjectivity, authorship, and reflectivity. *Counselling Psychology Quarterly, 25*, 347–360.

Suoninen, E., & Wahlström, J. (2009). Interactional positions and the production of identities: Negotiating fatherhood in family therapy talk. *Communication & Medicine, 6*, 199–209.

Wahlström, J. (1990). Conversations on contexts and meanings: On understanding therapeutic change from a contextual viewpoint. *Contemporary Family Therapy, 12*, 455–466.

Wahlström, J. (2006a). The narrative metaphor and the quest for integration in psychotherapy. In E. O'Leary & M. Murphy (Eds.), *New approaches to integration in psychotherapy* (pp. 38–49). London: Routledge.

Wahlström, J. (2006b). Narrative transformations and externalizing talk in a reflecting team consultation. *Qualitative Social Work, 5*, 313–332.

Wetherell, M. (2001). *Discourse as data: A guide for analysis*. London: Sage.

Widdicombe, S. (1998). 'But you don't class yourself': The interactional management of membership and non-membership. In C. Antaki & S. Widdicombe (Eds.), *Identities in talk* (pp. 52–70). London: Sage.

Chapter 11
About Complexity, Difference, and Process: Towards Integration and Temporary Closure

Maria Borcsa and Peter Rober

Therapists and clients are in language and they make use of language. A couple, a family, try to understand each other (literally and metaphorically) just like therapists do in their exchanges with their clients. This is a process in time: meaning, rhetorical figurations, structural patterns, etc. develop in sequences. How change is induced through and during these sequences is a question of special interest for psychotherapy research (Elliott, 2012). However, from a systemic point of view, every communicative intervention, be it verbal or nonverbal, can only be understood as a perturbation (Luhmann, 2012) in our clients: in therapeutic change via communication, there is no unilateral causality. Language as such, with its potential to create diverse implications, is manifold and generous. Therefore, we can ask: how come that individuals, couples, families but also therapists "choose" one meaning—by reacting to one option—and don't choose others? This is a question qualitative researchers are interested in: in what way can the data be understood as meaningful rather than happenstance? How can we grasp the necessity of what we have observed?

This Book

This book has presented a variety of ways to deal with these questions and the challenge we outlined in Chap. 1: the challenge of qualitatively researching multi-actor therapeutic sessions with respect for the specificity of the setting.

M. Borcsa (✉) • P. Rober
Institute of Social Medicine, Rehabilitation Sciences and Healthcare Research, University of Applied Sciences Nordhausen, Nordhausen, Germany
e-mail: borcsa@hs-nordhausen.de

© Springer International Publishing Switzerland 2016
M. Borcsa, P. Rober (eds.), *Research Perspectives in Couple Therapy*,
DOI 10.1007/978-3-319-23306-2_11

Each chapter is a (re-)construction of the story of Victoria and Alfonso and their therapist(s), and each re-telling differs from the next one. However, what all approaches have in common is the meticulous and careful reading of the transcripts and a commitment to the rich meanings implied in the interlocutors' words. Because of this commitment and the complexity of multi-actor dialogues, the researchers had to choose a lens through which they wanted to look at the sessions; they had to select a focus. These foci were more methodological (e.g., discourse analysis, objective hermeneutics), conceptual (e.g., power, moral order), or related to the material (e.g., the start of the first session, the beginnings of all four sessions). These choices had to be made in order to set the stage for the actual research: the careful retrospective analysis of what happened in the four sessions with Victoria and Alfonso.

Each author explains how he/she understands what happened with the couple in the therapy. They back up their interpretations by the data, and refer time and again to the transcripts, pointing to certain interactions or words that were spoken. In qualitative research, the reliability and trustworthiness of the author's interpretation are also left to the reader, who can judge for him/herself if the way the author understood the words of the interlocutors makes sense in the context.

While different choices were made by the different authors, no one claims to own the truth. Throughout the book, there is an awareness of the complexity of multi-actor sessions that precludes claims of some methodologies being right while others are allegedly wrong. All authors know full well that their approach, while viable, is just one among many valid and useful approaches. The multiplicity in this book testifies to richness and creativity, rather than to competition or rivalry.

Victoria and Alfonso: Some Final Remarks

At a certain stage of a couple's relationship and during the first therapy session(s), the partakers may consider certain topics to be too challenging for the autopoiesis (Maturana & Varela, 1980) of the system. Or, to put it differently: not to risk the breaking apart of the (therapeutic or even couple) relationship, social actors decide more or less consciously not to speak about certain topics during a certain period of their interaction with others. In clinical practice, we all know how multifarious possible turn-takings at the very beginning of the first session can be and we take precautions not to close too quickly the many possibilities language provides us with: the therapist in the case of Victoria and Alfonso knows, on the one hand, about the expectations with regard to his role and that he must give a certain structure to the communication. Nevertheless by asking "Where would you start?" he gives the maximal amount of freedom to his communicative partners and, in doing so, we may say: he starts without starting.

In young couples, like in the case of Victoria and Alfonso, we can witness certain negotiations which are conducted as a developmental task of the couple's system. One of these aspects is the question of loyalty to the family of origin of the respective partner (Boszormenyi-Nagy & Spark, 1973). The positions each person holds in relation to Alfonso's family of origin are ambiguous (as Victoria's parents are not much talked about, these relations are out of focus). Therefore, the couple system can be seen as receptive to therapeutic interventions on this topic. In our case these interventions are clarifying, supportive (especially for Victoria), as well as challenging (especially for Alfonso). The therapist gains a certain power of influencing this developmental task as a socializing agent himself, a structural aspect which can be reflected on, but not denied (Guilfoyle, 2003).

The case of Victoria and Alfonso is noteworthy, as it brings together aspects of an intercultural couple, as well as aspects of a couple where one person is diagnosed with depression. Partners who are approaching a relationship from different cultural backgrounds have an additional task to solve, which is to create a way of integrating their different cultural experiences in their life as a couple. They have at least four ways of dealing with this challenge (Seshadri & Knudson-Martin, 2013): (1) they can blend both cultures and rejoice at each; (2) they can coexist: they view their cultural differences as positive and attractive and retain them separately; (3) further, one partner may be more assimilated to the partner's culture than the other one, while this solution is not regarded as a compromise but as the right way of dealing with this issue by both sides. Possible risk groups are (4) couples who do not know what to do with the differences; they might ignore them, which creates insecurities in the relationship.

It seems obvious that the case of Victoria and Alfonso belongs to the last group, uncertain and unstable in their dealing with their cultural differences. As they happen to live in Victoria's home country, she expresses a need to stick to her ideas of a good home. This is difficult for Alfonso. Still, one gets the impression that during the sessions a change manifests in Alfonso, who appears to become more tolerant to cultural differences in his everyday life (e.g., having carpets in the new apartment: yes or no?). As we have no possibilities of a follow-up session, we do not know if this tolerance was more an expression of temporary conflict avoidance from Alfonso's side or a sustainable pattern of positive singular assimilation to the new "home" country.

Subjective experiences of feeling insecure play an important role in Victoria's narration about her being depressed; this is mirrored in research results of interactional patterns of couples burdened with depression (Beach, Dreifuss, Franklin, Kamen, & Gabriel, 2008). Meta-analyses show a highly significant cross-sectional and processual association between couple distress on the one hand and depression on the other (Beach & Whisman, 2012). Humiliating events can be severe stressors and increase the risk of showing symptoms, while personal biographical vulnerabilities (connected to one's own family of origin) may be part of a larger vicious circle (ibid.). It seems that in the case of Victoria and Alfonso, the following aspects are intertwined: a biography of subjectively felt insecurity (Victoria: "I have these issues with trust," fourth session) as a personal vulnerability, triggered by a humiliating event of being "rejected" by Alfonso's family of origin, where she is not able to communicate and feels helpless. The powerlessness increases as Alfonso is not following her wishes to stay in close—even if mediated—contact. Interestingly,

Victoria's fantasized rivals are not other women in Alfonso's home country but his family of origin. She is not afraid that Alfonso connects with another woman but that he cannot disconnect from his mother and his original family (to be remembered: he is the youngest son) to create his own. This could relate to the perspective Victoria may have on their different positions in their respective life cycles: while she was already working before starting her studies, and works in parallel with her studies to make her living, it seems that Alfonso relies financially on his parents and an international exchange grant. While she is speaking about having children together, he is not expressing a clear commitment to Victoria as a life partner. He may be seen as a representative of the new phenomenon of the "in-between age," perhaps not consciously deciding to be irresolute but being influenced by broader social changes (Arnett, 2000; Chisholm & Hurrelmann, 1995).

Mirroring these dynamics, no attraction to Alfonso's culture is reflected in Victoria's narrations—a *positive* coexistence of the two cultures is not visible at this point of their partnership. An integrated way of dealing with this issue, where both cultures are conjointly "celebrated" seems out of sight: the next Christmas will be spent separately in their two different countries. Nevertheless, Victoria acknowledges the change in Alfonso's contact behavior during his last stay in his country of origin at the end of therapy: he used the mobile phone more often to show commitment (Bacigalupe & Cámara, 2012; Bacigalupe & Lambe, 2011).

Therapy Process and Therapist's Role

All authors in this book have dealt with the complexity of multi-actor therapeutic settings. Some have conceptualized this complexity in systemic ways emphasizing the existence of implicit meaning structures in the couple that can explain interactional patterns. Ugazio & Fellin have presented a semantic approach which explains problems in couples as a result of a semantic mismatch creating misunderstandings and conflicts. The aim of therapy then is to contribute to a greater semantic cohesion as a foundation for trust and positive communication. Borcsa presents a hermeneutical approach. When the couple's complexity is described in terms of interactional patterns based on latent meaning structures, change in these nonfunctional patterns is the target of psychotherapeutic interventions in couple therapy. Therapy then is not only focused on the disruption of these patterns but also on changing the couple's structural rules.

Other authors have preferred to use a more discursive or narrative frame to approach the presented case. Avdi casts a closer look on how talk can be therapeutic. She reconstructs the contributions and the discursive agenda of the therapist, pointing at the fact that the therapist's turns fulfil a function. Wahlström's discursive analysis is bound to the viewpoint of a couple's relationship as a social institution with a particular moral order. Päivinen & Holma describe therapy as a process of storytelling in which the therapist directs the couples' narrations. Reflexivity and acknowledgement of power issues are necessary in order to bring marginal stories to the forefront as results of therapeutic interventions.

Several authors, like Rober, Seikkula & Olson, and Laitila have used a dialogical perspective in that the process of languaging is central. For these authors, the therapist's role seems to be limited to being responsive and favoring a therapeutic presence of empathic listening and a courageous attitude towards exploring the details of the participants' speech. The role of the therapist seems to be to open space for new ways of speaking and to assist the clients in developing a new language. Change can be described as the emergence of new and different voices that allow for the possibility of a real dialogue. Rober has described how such new language emerges in the session, using the central concept of positioning. He changes the figure and background by introducing the model of "dialogical space." This space is the one in which clients and therapists create dialogues as a dance of invitations and responses. The dialogical entitlements evolve in the process of therapy.

While the authors in this book who adopt a dialogical perspective defend the modesty of the therapist's role as a responsive, non-interventionist presence, this viewpoint is problematized especially by Wahlström. He asks significant questions: should therapists take up a problem as formulated by the couple itself and hence confine themselves to the scope of solution such a formulation might imply? Or should therapists rather use their knowledge and experience to address issues that are not explicitly mentioned by their clients as being central to their struggles?

Here we see how very much the stance of the therapist contributes to his own positioning in the multi-actor therapeutic setting.

Towards a Temporary Conclusion

The different authors in this book approached the case of Victoria and Alfonso through theories of language (Bakhtin, 1981; Harré & Van Langenhove, 1999), social theories (Berger & Luckmann, 1966; Foucault, 1977; Goffman, 1959), systems theories (Bertalanffy, 1968; Luhmann, 2012; Maturana & Varela, 1980), or clinical theories (Minuchin, 1974; Ugazio, 2013; White & Epston, 1990). From each perspective the emphasis is slightly different: we can perceive therapy sessions as a dialogical realization of relatedness through language, as an arena for certain social negotiations, as a field of (re-)production of a system's patterns, or as a portrayal of a representative clinical case.

These possibilities are provided by the material itself, as the researchers in this book applied no specific research methods *to collect* the data (see, e.g., Elliott, 2012). This book's general tenor of using simply videotaped and transcribed therapy sessions as "naturally occurring talk" is reflected in the freedom of the variability of methodologies used by the authors. The different approaches shed light on various phenomena and they extend beyond psychotherapy research in the narrow sense, especially when they focus on concepts like "moral order" (Chap. 10), or "power" (Chap. 7). These concepts invite more general descriptions of social systems, relevant not only to clinical cases but to all couples, families, and other social systems. Here we see that in working systemically (as researchers and practitioners) we do—and have to—consider characteristics of social life at large; working with couples or families always entails working with structural aspects of mankind as social and discursive beings as well.

This book is living proof of the value of qualitative research for addressing the complex interactions that therapeutic multi-actor sessions are. The different research approaches and tools presented illustrate only some perspectives we can take on couple therapy. They demonstrate the usefulness of studying therapeutic sessions in their specificity in order to deepen our understanding of the conversational processes we call "couple therapy."

References

Arnett, J. J. (2000). Emerging adulthood. A theory of development from the late teens through the twenties. *American Psychologist, 55*(5), 469–480.

Bacigalupe, G., & Cámara, M. (2012). Transnational families and social technologies: Reassessing immigration psychology. *Journal of Ethnic and Migration Studies, 38*(9), 1425–1438.

Bacigalupe, G., & Lambe, S. (2011). Virtualizing intimacy: Information communication. technologies and transnational families in therapy. *Family Process, 50*(1), 12–26.

Bakhtin, M. (1981). *The dialogic imagination.* Austin, TX: University of Texas Press.

Beach, S. R. H., Dreifuss, J. A., Franklin, K. J., Kamen, C., & Gabriel, B. (2008). Couple therapy and the treatment of depression. In A. S. Gurman (Ed.), *Clinical handbook of couple therapy* (4th ed., pp. 545–566). New York: Guilford.

Beach, S. R. H., & Whisman, M. A. (2012). Affective disorders. *Journal of Marital and Family Therapy, 38*(1), 201–219.

Berger, P., & Luckmann, T. (1966). *The social construction of reality. A treatise in the sociology of knowledge.* Garden City, NY: Anchor Books.

Bertalanffy, L. von (1968). *General system theory: Foundations, development, applications.* New York: Braziller.

Boszormenyi-Nagy, I., & Spark, G. (1973). *Invisible loyalties.* New York: Harper & Row.

Chisholm, L., & Hurrelmann, K. (1995). Adolescence in modern Europe: Pluralized transition patterns and their implications for personal and social risks. *Journal of Adolescence, 18*, 129–158.

Elliott, R. (2012). Qualitative methods for studying psychotherapy change processes. In D. Harper & A. R. Thompson (Eds.), *Qualitative research methods in mental health and psychotherapy. A guide for students and practitioners* (pp. 69–81). Chichester, UK: Wiley.

Foucault, M. (1977). *Discipline and punish: The birth of the prison.* London: Penguin Books.

Goffman, E. (1959). *The presentation of self in everyday life.* New York: Doubleday Anchor.

Guilfoyle, M. (2003). Dialogue and power: A critical analysis of power in dialogical therapy. *Family Process, 42*(3), 331–343.

Harré, R., & Van Langenhove, L. (Eds.). (1999). *Positioning theory.* Oxford: Blackwell.

Luhmann, N. (2012). *Introduction in system's theory.* Chichester: Wiley.

Maturana, H., & Varela, F. (1980). *Autopoiesis and cognition: The realization of the living.* Dordrecht: D. Reidel.

Minuchin, S. (1974). *Families and family therapy.* London: Tavistock.

Seshadri, G., & Knudson-Martin, C. (2013). How couples manage interracial and intercultural differences: Implications for clinical practice. *Journal of Marital and Family Therapy, 39*(1), 43–58.

Ugazio, V. (2013). *Semantic polarities and psychopathologies in the family: Permitted and forbidden stories.* New York: Routledge.

White, M., & Epston, D. (1990). *Narrative means to therapeutic ends.* New York: Norton.

Index